Radio's America

RADIO'S AMERICA

The Great Depression
and the Rise of Modern Mass Culture

BRUCE LENTHALL

The University of Chicago Press · Chicago and London

Bruce Lenthall is director of the Center for Teaching and Learning and an adjunct assistant professor in the Department of History at the University of Pennsylvania.

The University of Chicago Press, Chicago 60637
The University of Chicago Press, Ltd., London
© 2007 by The University of Chicago
All rights reserved. Published 2007
Printed in the United States of America
16 15 14 13 12 11 10 09 08 07   1 2 3 4 5

ISBN-13: 978-0-226-47191-4 (cloth)
ISBN-13: 978-0-226-47192-1 (paper)
ISBN-10: 0-226-47191-8 (cloth)
ISBN-10: 0-226-47192-6 (paper)

Library of Congress Cataloging-in-Publication Data

Lenthall, Bruce.
    Radio's America : the Great Depression and the rise of modern mass culture /
Bruce Lenthall.
        p. cm.
    Includes bibliographical references and index.
    ISBN-13: 978-0-226-47191-4 (cloth : alk. paper)
    ISBN-10: 0-226-47191-8 (cloth : alk. paper)
    ISBN-13: 978-0-226-47192-1 (pbk. : alk. paper)
    ISBN-10: 0-226-47192-6 (pbk. : alk. paper)
    1. Radio broadcasting—United States—History. 2. Radio broadcasting—Social
aspects—United States. I. Title.
    PN1991.3.U6L46 2007
    302.23'44097309043—dc22

                                                                2007007194

♾ The paper used in this publication meets the minimum requirements of the American
National Standard for Information Sciences—Permanence of Paper for Printed Library
Materials, ANSI Z39.48—1992.

For Calista, Chapin, and Lia

CONTENTS

ACKNOWLEDGMENTS

The people who have offered me their guidance or support over the years that I have lived with this project deserve and have my thanks.

As this project developed as a dissertation, Bruce Kuklick was an ideal advisor; the relationship is less formal but no less welcome now. Tom Sugrue, Peter Conn, and Barbara Savage all provided much-appreciated guidance. Discussing my work and the process of research and writing with Beth Clement and Randolph Scully was also particularly helpful. More recently, Josh Frost, John Noakes, and Eric Schneider contributed valuable—and timely—comments and conversations.

In revising and developing my arguments, I have benefited immensely from teaching related topics. My Twentieth-Century U.S. Cultural History classes and my Media in American History classes at Bryn Mawr College and Barnard College significantly improved this book. At the University of Chicago Press, Douglas Mitchell's shepherding of this project and Carol Saller's thoughtful editing have been welcome. An earlier version of a portion of chapter 1 appeared as "Critical Reception: Public Intellectuals Decry Depression-era Radio, Mass Culture, and Modern America" in *Radio Reader: Essays in the Cultural History of Radio* (2002), and is reproduced by permission of Routledge/Taylor and Francis Group.

Writing this book I have been especially grateful for the support my family has given me. My mother, Pat Riley, my father,

Jerry Lenthall, my brother, sister, and sister-in-law—Josh, Lisa, and Melanie Lenthall—have all helped me to meet the challenges I encountered. Finally, let me voice my appreciation for Calista Cleary, Chapin Lenthall-Cleary and Lia Lenthall-Cleary. Calista deserves my effusive thanks: she has read, commented on, and discussed this work again and again and again; far more than that, she has been one of the best parts of my life. Although few of their contributions have much to do with this book, I have learned more from Chapin and Lia than I ever imagined; I appreciate what they mean to me beyond words.

When I cast my mind back to the summer of 1936 different kinds of memories offer themselves to me. We got our first wireless set that summer—well, a sort of a set; and it obsessed us. . . . I remember my delight, indeed my awe, at the sheer magic of that radio. And when I remember the kitchen throbbing with the beat of Irish dance music beamed to us all the way from Dublin, and my mother and her sisters suddenly catching hands and dancing a spontaneous step-dance and laughing—screaming!—like excited school girls, at the same time I see that forlorn figure of Father Jack shuffling from room to room as if he were searching for something but couldn't remember what. And even though I was only a child of seven at the time I know I had a sense of unease, some awareness of a widening breach between what seemed to be and what was, of things changing too quickly before my eyes. . . .

—Brian Friel, *Dancing at Lughnasa*

INTRODUCTION

"The Story of the Century"

Listeners who tuned into the radio Sunday, October 30, 1938, shortly after 8:00 p.m. heard an announcer interrupt the music of Ramon Raquello's orchestra for a special news bulletin. Several astronomers, the announcer declared, had observed inexplicable and fiery explosions on the surface of the planet Mars. The announcer quickly returned listeners to the orchestra, but promised to cut away again to take listeners to an interview with a noted astronomer as soon as possible. A few moments later, the announcer introduced a reporter and professor Richard Pierson. "Please tell our radio audience exactly what you see as you observe the planet Mars," the reporter asked. The voice of Orson Welles answered. And with the actor's reply, the *Mercury Theater on the Air*'s broadcast of *War of the Worlds* sprang out of its windup.[1]

Over the next hour, Welles and his troupe dramatized the near destruction of the world as attacking Martians overran humanity. Welles's account of an event that never took place—the invasion from Mars—would become one of the most renowned single radio broadcasts ever. Listeners who tuned to CBS that Halloween eve heard Martians land spacecraft in New Jersey and around the country. Those listeners heard the Martians annihilate the populace with heat rays and lay waste to the military with poison gas. They heard warning bells toll and radio communications falter as the invaders overcame New York City. And in the end, listeners heard the mighty Martians vanquished by bacteria for which their alien immune systems had no defenses.[2] By that point, however, a sizable minority of listeners had long since stopped listening. They were too busy panicking.

Over one-million listeners, about one-fifth of the program's audience, believed the broadcast invasion to be both real and terrifying. The program's enduring fame, of course, comes from the fear it spawned. Listeners fled, clogged phone lines seeking information, prayed, went into shock, and contemplated suicide rather than die at the Martians' hands. The crowds that flocked New York City's streets, said one observer, outdid even the chaotic scene that had accompanied the end of World War I. A New York man wished for a gun so that he and his family could take their own lives; lacking a weapon, he instead packed hurriedly and called friends warning them to flee as well. A Tennessee woman spent the evening praying on her kitchen floor, and the host of a California party reported two of his guests suffered heart attacks when they heard of the invasion. In Trenton, New Jersey, the broadcast crippled city communications as two thousand callers phoned the police department in two hours; fortunately there was no real emergency, the city manager said, because with all municipal lines tied up, there would have been no way to dispatch firefighters and the like. How widespread would such hysteria have been had *Mercury Theater on the Air* drawn more than a tiny share of the radio audience that Sunday night?[3]

Only a scattering of Americans, however, even heard the drama that would become so well remembered as an example of radio during its heyday. Welles's program was not a popular one. CBS broadcast *Mercury Theater on the Air* opposite the most widely listened-to program on radio in 1938, the *Edgar Bergen–Charlie McCarthy Show*. No advertiser saw fit to sponsor the Mercury Theater's radio offerings. In a nation of roughly 130 million people, only approximately 6 million even heard Welles's broadcast; far fewer, of course, found it scary. We might be tempted, then, to dismiss the program's resonance as an ironic twist of memory: we often recall the unusual and rare, not the commonplace and typical.

But there is more to the story of the *War of the Worlds* broadcast than a tale of space invaders and ensuing panic. Welles's program might not have been widely heard and it might have provoked an unusually dramatic reaction in only a few, but the broadcast and surrounding events reveal Americans integrating the new medium of radio into their lives in the decade of the Great Depression. In the 1930s a new system of mass communications took hold in the United States and helped to spawn a new mass culture. As radio brought an expanding, impersonal public sphere home to Americans, they encountered a world in which even culture and communication might be centralized and standardized. The modern culture that radio represented threatened to overpower individuals, leaving them with little control either in their own lives or in the wider world. As public intellectuals of the day lamented, that culture might be as menacing as Welles's Martians.

But mass culture did not quite prove entirely all-powerful. Within very constrained space, Americans found some room to interpret radio's meanings for themselves. As they did so, Americans used those meanings to help them address the very challenges posed by radio and mass culture more generally. The story of Welles's unusual broadcast resonated with these more typical experiences and understandings of Americans seeking to come to terms with radio and the rising culture it represented. Many found radio could enable them to gain a sense of autonomy in their own lives by helping them understand an encroaching mass world in familiar, personal terms. To some, radio also offered the prospect of speaking meaningfully in that world, through a newly viable hope of communication with a mass audience. As Americans determined what radio meant to them, they used the mass medium itself to gain a measure of control and perhaps a voice within the disempowering mass world that radio helped to create. And this process, in turn, helped to shape that modern world.

Embedded in the story of the Mercury production, then, lie the shades of the twentieth century's rising mass culture, which, as public intellectuals in the 1930s rightly noted, threatened to erode individual distinction and choke off individual voices. To such thinkers, Welles's program must have seemed to affirm their fears. In the reaction to the radio play, journalist Dorothy Thompson saw Americans losing the ability to think on their own and falling prey to an insidious homogeneous thought. The broadcast, she wrote, "proved how easy it is to start a mass delusion." To Thompson, radio could help engender a new mass mind. To other intellectuals, the mass media narrowed control of public speech. With Welles's newfound fame, his program attracted a corporate sponsor for the first time; the *Mercury Theater on the Air* became the *Campbell Playhouse*. As Welles learned, sponsorship came with a price. Welles wanted to spread high culture, to make political statements, to experiment with his art; Campbell's wanted to maximize their audience and to sell soup. In the end, the soup sellers, not Welles, had the final say on the program.[4] Radio, Welles might have concluded and many public intellectuals believed, embodied the dangers the fledgling century posed to individuals' autonomy in their own lives and abilities to make a public impact.

To many critics, including Thompson, it came as a surprise that so many ordinary listeners took the *War of the Worlds* program literally. It need not have. Listeners in the 1930s came to rely upon radio. True, *War of the Worlds* was only one particularly disquieting example of the way in which radio caught Americans up in a far-off and confusing world. But at the same time, radio could familiarize that mass world. Listeners formed imagined but meaningful relationships with radio voices. Through radio, listeners remade the frightening public sphere in comfortable and comprehensible private terms. Listeners taken in by Welles's hoax often suggested indignantly that they had believed in the startling invasion because they had learned—appropriately, many felt—to have faith in radio. Of course they accepted what they heard, they declared: radio helped them stay afloat amidst waves of change they felt breaking around them. "In this troubled world of ours," wrote F. M. Moody of Venice, California, explaining why he believed Welles's broadcast, "there are so many things that have happened and are happening that, we the people are believing nothing is impossible and naturally rely on our radio to bring correct, unadulterated, authentic, happenings of the world."[5]

The private connections listeners felt with public voices on the air and the resources listeners derived from those relationships had specific implications for the practice of democracy in a mass society. Told through a series of newscasts, Welles's radio play and the fact that listeners obviously incorporated those newscasts into their lives suggested that radio had the capacity to inform listeners and involve them directly in national civic affairs—for many, for the first time. To many listeners, the personal reach of radio gave them a way to count in a political arena that, without broadcasting, felt overwhelming. This capacity, many contemporaries felt, suggested radio might reinvent democracy to make it meaningful in their extended society.

At the same time, though, the *War of the Worlds* broadcast warned that the process of reimagining democracy for a mass society was, at the very least, complex—dangerously so. With the rise of broadcasting, as the possibility of speaking to the whole of a nation at once became more and more real, the meaning of communication changed. Audiences seeking access to influential voices welcomed this development, but Welles's program clearly revealed the thin line between demagoguery and democracy in an era of mass communication: broadcasting made it possible for a single powerful speaker to move a large part of the country at once. Despite his protestations of innocence at the time, Welles may have intended to make that very point with his broadcast. In a 1955 interview, Welles explained the famous program in similar terms: "We were fed up with the way in which everything that came over this magic box, the radio, was being swallowed," he said; radio in the 1930s "was a voice of authority. Too much so. . . . It was time for someone to take some of the starch . . . out of some of that authority: hence my broadcast."[6] Intended or not, certainly the broadcast contained a warning about the mass media's real and ominous potential to concentrate power by amplifying particular voices.

For others, however, this new kind of communication—communication that enabled a speaker to reach an enormous collective audience—suggested more benevolent possibilities. Mass communication, some felt, might enable individuals to find voices meaningful in the twentieth century's vast national public sphere. To a team of social scientists from the influential Princeton Radio Project, the Mercury broadcast indicated radio's promise as a means of communication for a far-flung but interconnected society. Radio Project researchers responded to the program by conducting an investigation into the panic that followed. The study, claimed lead author and psychologist Hadley Cantril, might reveal how to use radio to communicate with the public at large in order to better manage the problems of a modern society. Making sense of radio and the panic Welles had induced, Cantril suggested, could help experts take advantage of the newfound prospect of speaking to the masses that radio might provide.[7]

Indeed, Welles himself relished the possibility of finding a voice that might be heard by a vast populace. In a mass society, truly relevant communication had to speak to a mass public, Welles and a cadre of new radio writers and directors believed. Consequently, the mass communication radio made possible might be essential. Radio could revive the significance of an individual's speech and, more particularly for this artistically ambitious group, art. After all, while many Americans looked upon the Halloween-eve program as a cunning trick, to its creators and a sizable body of listeners, the broadcast offered an artistic treat. The *Mercury Theater on the Air*, raved New York City listeners Beaumont and Nancy Newhall, "made an original contribution to the difficult and unexplored medium of radio. By using radio's own technique, [Welles] made superficially believable the entirely unbelievable fantasy of H. G. Wells." The Newhalls had nothing but praise for the drama's "courageous and unceasing experimentation in America's most original and promising art-form—radio."[8] Not just a holiday stunt, the *War of the Worlds* production was part of an attempt to use radio to create a new artistic voice that would be meaningful in the modern world.

Clearly, as Americans responded to the Mercury broadcast of the fictional disaster, they simultaneously expressed their reactions to far more real cultural disruptions. During the Great Depression, many Americans first found their lives tied into an unfamiliar, vast and abstract world. And during the Great Depression, many Americans began figuring out how they would inhabit that world. Radio, the 1930s and, more broadly, the twentieth century transformed the United States. But at the same time, Americans found in radio itself means to address the very challenges their new mass culture posed.

As Americans integrated broadcasting into their lives for the first time, they practiced a precarious balancing act. In the 1930s—and beyond—critics of mass culture came to see it as dangerously monolithic, as imposing one set of thoughts and values. In the 1930s—and beyond—celebrants of mass culture came to see it as a populist marketplace, a forum in which the tastes of the majority ruled. Neither, in fact, got it fully right. Radio embodied the new centralized and standardized mass culture beginning to take hold in the United States in the Depression decade. The critics were right that radio and mass culture did constrict the choices audiences could make. But Americans did not fully accept the standardized meanings—of radio or their mass culture more generally. Within tight limits, they found some of their own.

By the Great Depression, as the country became increasingly interconnected, Americans found their worlds becoming larger and larger—and their own places within those worlds becoming smaller and smaller. The culture of the twentieth century would be, in important respects, mass-produced by a few and designed for mass consumption by a wide sweep of the nation. And the revolution in communications that radio represented played an essential role in this development. The mass society of the twentieth century first coalesced in the 1930s, and to the many Americans who had previously been apt to see personal and local experiences as most central to them, the sense of belonging to a mass society hit painfully. The new culture threatened to strip Americans of meaningful power both in their own lives and in the broader public arena. How could you retain your self-control when an overwhelming outside world increasingly intruded into and dictated your daily life? How could you make your own marks on such a colossal and distant sphere?

In the face of such problems, Americans often found they could draw upon the leading purveyor of their new culture: radio. As Americans used radio to help them make their mass world personal, its intrusions no longer felt so disempowering, and the possibility of counting in that world no longer seemed so impossible. By making mass relationships resemble private ones, listeners gained a sense of control in their own lives and a sense of standing in the sprawling public arena. To some, broadcasting might also enable an individual to be heard in the seemingly ever-widening world. They had come to believe in the possibility of—and need for—genuine communication with a collective audience. No matter how constrained Americans' choices were as they made sense of radio, those choices mattered. The meanings Americans found in radio provided them with ways of navigating their new mass world, helping to shape how we would live in that world and, in turn, that world itself.

As Americans incorporated broadcasting into their lives and found a sense of autonomy and perhaps voice in their new mass culture, they engaged in a process that generated powerful changes. On some level, Americans came to believe that a common public existed and that it was possible to address that mass at large—while connecting with the members of that mass on a personal level. This is essential. It would blur the divide between public and private, and revise the meanings of democracy and communication itself.

In the broadcasting age, listeners came to imagine they could enter the public world in private terms. Radio brought far off voices and events into the home in a seemingly intimate fashion. This eroded lines between the public and private spheres, making it more difficult to distinguish public and private and to determine separate priorities and behaviors for each.

This, of course, had consequences for democracy. If and how citizens could count in the public sphere and the very idea of democracy in the modern age were up for debate in the 1930s. The interconnected and vast society of the twentieth century made older notions of participatory and local democracy seem nonsensical. With the rise of broadcasting, instead the idea of democracy often became embodied in feelings of quasi-private relationships with public figures and in the idea of a politically informed populace, not necessarily in personal involvement in decision making or public action.

All of this only made sense because of new understandings of communication. In the 1930s, many Americans were coming to accept the notion of mass communication. Ideas as to what constituted genuine communication expanded: rather than simply an interchange between individuals, many believed authentic communication could include one-way transmissions to an impersonal, vast audience. That is, for many in the United States the meaning of "to communicate" came to emphasize "to make common" more than "to share" or "to exchange." Today we may accept these notions so easily that it is hard to imagine a moment when this—any of it—was controversial and, to some, alarming. In fact, though, the arrival of a mass culture and radio early in the twentieth century and in the Depression era in particular helped to transform the meanings of public and private, of democracy, of communication, giving birth to modern versions of these ideas and practices.

The mass culture that took hold in the United States around the 1930s was neither entirely oppressive nor particularly flexible. But it pushed real changes. As Americans negotiated some of their own cultural meanings within constrained limits, they helped define the meaning of radio, mass culture, and the modern United States. In doing so, they struck a fine balance. These balances, the meanings Americans found in radio, had some very real repercussions in the Depression and beyond. And those repercussions involved trade-offs, some positive and some negative elements. We cannot and should not be entirely comfortable with the changes mass culture delivered. The sense of autonomy that Americans gained relied, ultimately, on an element of illusion: the personal bonds so many felt were spun of nothing more than air. And the concept of public participation frayed. Yet even so, those slender threads made real differences to many struggling with the unsettling realities of their emerging mass world. How Americans began to find their way in that world was, in many respects, the story of the Depression. And more than that: the growth of a modern mass culture—and how Americans dealt with the tensions it spawned—would compose a large part of the story of the United States through the twentieth century.

That modern culture differed markedly from what came before. The world of the twentieth century that would seem so familiar, perhaps almost natural, decades later was, in fact, a disconcertingly new creation when it arrived. Starting in the late 1800s, but only fully arriving in the Great Depression, modern mass culture—an increasingly vast, standardized, centralized, and impersonal world—gradually emerged. Through the late nineteenth century and into the twentieth, the fibers linking the nation twined into thicker and thicker cords, expanding the scope of an individual's world. People and communities formerly oriented locally increasingly found themselves enmeshed in arenas that extended far beyond traditional face-to-face experiences. The dramatic expansion of the railroads in the decades from the Civil War to World War I, for instance, brought far-flung locales more firmly into a national economic web. The boundaries of space softened. By the end of the 1920s, many people who only a few decades earlier would have only rarely traveled beyond walking distance had access to cars or other mechanical transportation. Moreover, urbanization and the rise of large, bureaucratic institutions created new social relationships. As cities swelled—over 500 percent from 1870 to 1920—many Americans came into contact with a wider range of people than ever before. And as nationwide organizations—from reform groups to corporations to professional organizations and unions—took on more prominent roles, many Americans discovered that some of their crucial bonds spanned nineteenth-century geographic limits. Expanding communications facilitated and represented this process. Around the turn of the century, the post office first began home delivery of mail to rural addresses; within thirty years, use of first-class mail increased 70 percent. More dramatically, with the late-nineteenth-century popularization of the telegraph, news reports were distributed quickly across space, tying many into national events.

As part of that course, American society began a process of standardization of sorts. The ongoing development of mass-production techniques meant that, more and more, Americans across the country could own the same goods. The Sears catalogue of the late nineteenth century and the chain stores of the twentieth century might carry the same sewing machines or soap to widely distributed consumers. And as producers, industrial workers and professionals alike more and more often labored under uniform work guidelines set by central authorities, from factory officials to professional associations. It was not just objects that were mass-produced. Across the country, Americans could increasingly encounter similar values, information, and entertainment. Businesses and advertisers early in the twentieth century began seeking ways to inspire mass consumption. That meant finding or creating and then emphasizing what large groups of consumers had in common. Similarly, the growing newspaper wire services and chains made it more likely that readers would receive the same information, with perhaps the same slant, regardless of locale. And as movies developed early in the twentieth century, audiences from the Bay State to the Bay Area had easier and easier access to the same mass-produced entertainment, offering, theoretically, the same messages.

Inherent in this process was a measure of centralization of social, economic, and cultural authority. More and more, influential decisions were made at a distance from ordinary lives. It would be easy to misstate the standardization taking place as one that produced an unmitigated sameness throughout life. The turn-of-the-century decades, in fact, also witnessed rising variety in many areas: tremendous waves of immigration and migration generated new levels of ethnic diversity, for instance. But many aspects of American culture did become more homogeneous and many forms of diversity were constrained as particular centralized powers found new capacities to distribute uniform products, ideas, and practices to the populace as a whole. The trend of the era pushed control of cultural, social, political, and economic life away from outlying individuals or communities and tended to centralize such authority. Americans slowly became more aware that they were part of a national network in which important economic and political decisions or occurrences often took place far away, with comparatively few businesses, associations, or other institutions exercising rising influence. From the consolidation of corporations to the establishment of Progressive regulatory bodies to the organization of sports leagues, centralized power became increasingly visible and able to enact common standards for a widening swath of society. Again, new forms of communication—from newspaper chains to national advertising agencies to movie studios—only reinforced this narrowing authority. As they moved deeper into the twentieth century, Americans increasingly consumed news, advertisements, and entertainment produced by a limited number of distant authors.

All of this meant that slowly, ordinary Americans found the public arena expanding and growing more pervasive in their daily lives. As individuals became more and more intertwined in a lattice of people, institutions, and events beyond their own direct experiences, they found their interactions becoming more abstract and important relationships less personally based; they found meaningful access to the public sphere harder to come by. Growing bureaucratic structures—from a national economy to large-scale organizations—increasingly emphasized systemized interactions over interpersonal relationships. Conversations in the public world became less personal: newspapers increasingly replaced face-to-face contacts as sources of information, for instance; and those papers rarely depended on actual intimate ties between publisher and audience, but conveyed information to an abstract readership. These transitions were in their infancy, even by the 1920s. But they produced the beginnings of the culture that would become so familiar during the rest of the century. And many found being thrust into that vast world alienating. As the world most Americans inhabited grew, they came to find that traditional personal connections were decreasingly effective resources and that traditional ways of speaking in public no longer provided them with meaningful voices.[9]

These trends accelerated dramatically in the Great Depression. For many Americans, the 1930s brought the twentieth century's mass culture home. The crisis of the Depression was not only a sudden economic jolt, but the climax of disorienting cultural changes long in rising.[10] Nothing, of course, made clear to millions of Americans that they belonged to an interconnected and national economy like the devastating economic crisis. As local institutions—from banks to businesses to relief associations—failed, the Depression made plain the broad scope of even an individual's economic and social landscape. The decline of those local establishments along with personal hardship paved the way for many, particularly working-class Americans, to participate in centralized national institutions and the rising consumer culture. Chain stores and theaters, with standard products and selling practices, pushed aside many smaller, often ethnically diverse, shops during the Depression, for instance. In turn, the federal government's response to the adversity did more to centralize political power and to bring that distant authority into ordinary lives than any previous political endeavor. With programs from federal relief and regulation to rural electrification, the New Deal helped to make the central government an active part of many people's daily lives and consciousnesses. For many in the 1930s, coming face to face with a vast society—a world increasingly of far-off decision-making and impersonal, uniform interactions—proved confusing and devastating. It left them wondering how, or if, they could matter anymore. To author John Steinbeck that was the central crisis of the Depression. In the author's 1939 novel, *The Grapes of Wrath*, a tenant farmer pushed off his land by an abstract banking system demands to know how he can protect his family in such a world. "Who can we shoot?" he asks. In reply he hears only, "I don't know."[11]

A large part of the reason Americans felt these changes so deeply starting in the 1930s was, simply, radio. More than any preceding cultural vehicle, radio created and disseminated a mass culture. By the start of the Depression, after several decades of development, the national, commercially controlled, network system that would be the foundation of broadcasting in the United States was in place. And broadcasting revolutionized communication. Starting in the 1930s and continuing far beyond, the mass medium of broadcasting linked ordinary Americans to a widely dispersed world, centralized the authority to determine what programs would be conveyed to listeners throughout the country, and, in doing so, standardized the messages those listeners likely heard. Broadcasting delivered a massive public world into listeners' private realms.[12] Radio, as it developed in the 1930s, came to embody the culture of the twentieth century and first made that world a part of most Americans' daily lives, experiences, and ways of understanding.

In the early decades of radio, people around the globe marveled at what at the time seemed a truly amazing technology that could make possible communication through the air. It took several decades, though, to sort out the actual structure radio communications would take in the United States—a structure that would give rise to a culture both mass-produced and designed for mass consumption. Between 1897, when Guglielmo Marconi built on the work of other scientists and patented his wireless telegraph in England, and the 1920s, scientific advances came furiously and different interests struggled to shape the use of those advances. World War I jump-started a process of consolidating the control of radio in the hands of the government and a few businesses, but at that moment both those groups still conceived of the medium as a vehicle for point-to-point communication. Radio as a means of reaching a wide audience remained in the future; indeed, only the persistence of individual amateur radio operators revealed the possibilities in deliberately sending signals to a general audience.[13]

This practice of sending signals to a general audience—broadcasting—soon became the most prevalent expression of radio technology, of course. In the 1920s, an increasingly centralized system designed to capture a national audience developed. Corporate hands tightened their grip on radio and began airing programming designed to reach large audiences in order to turn a profit. In 1920, seeking a way to boost sales of its radio sets, the Westinghouse Corporation in Pittsburgh followed the lead of one of its employees, an amateur broadcaster: in November, with the broadcast of the presidential election returns, the station KDKA touched off an explosion of corporate stations with regularly scheduled broadcasting. By the end of 1922, the United States government had licensed 570 radio stations. In the mid-1920s, AT&T hit on a new way to fund broadcasting: the company initiated the practice of selling airtime on its New York station to other businesses. Such radio advertising caught on over the decade, thanks to the endorsement of the federal government. As the chief regulator of radio in the 1920s, secretary of commerce Herbert Hoover favored well-off and commercial stations—ones financially committed to maximizing their listenership—assigning them prize frequencies, for instance. In the wake of the Radio Act of 1927 that established the Federal Radio Commission, the new regulatory body followed suit, reallocating the nation's frequencies to commit the government formally to support commercial broadcasters and the advertising system. The National Broadcasting Company and the Columbia Broadcasting System took advantage of that climate, setting up the first national networks in late 1926 and late 1927 respectively. Seeking a mass audience, the two networks distributed programming to stations all across the county.[14]

As the Depression decade began, then, the modern American broadcasting system—dominated by centralized and commercial national networks—was newly up and running. The mass production of culture that radio made possible enabled a few broadcasters to blanket the nation. And with that, many Americans had to make sense of a new way of hearing the world for the first time. In 1926 no networks existed and fewer than 5 percent of all stations sustained themselves selling advertising. Roughly half a decade later the commercial giants NBC and CBS controlled 30 percent of the nation's stations—and those stations had such strength that the networks held roughly 90 percent of the broadcasting power. Sales of radio advertising, insignificant in 1927, topped a hundred million dollars in 1930. A few national advertisers, motivated by commercial goals, dictated the content of programming that listeners across the country tuned in to more and more regularly in the Depression. That opened space for some kinds of communication and threatened to drown out others. By the early 1930s the programming forms that would endure through radio's heyday—and eventually appear on television as well—took shape: situation comedies, variety programs, adventures, dramas, and political addresses to name a few. By the end of the decade, the two New York City–based networks had also become the leading sources of news for millions throughout the country.[15]

It was during the Depression that radio truly attracted a mass audience in the United States. As commercial enterprises, the networks created programs designed to be consumed by a wide following. And with the broadcasting system that would endure for most of the century in place, listeners tuned in to those programs in droves. Radio ownership more than doubled in the 1930s, from about 40 percent of families at the decade's start to nearly 90 percent ten years later. By 1940 more families had radios than had cars, telephones, electricity, or plumbing. To the members of that vast audience, radio, and particularly network offerings, became an integral part of their daily lives during the Depression. Americans loved the new medium, listening to each set an average of four to five hours a day. And with listeners preferring national programs to local ones by a ratio of nearly nine to one, audiences tuned those sets to network stations for most of those hours. Across the country, millions and millions of listeners heard the same programs for hours each day.[16]

Looking at this scenario in the 1930s, many public intellectuals reached the conclusion that America was fast becoming something new and frightening. Radio, these critics rightly asserted, helped to make a mass culture an everyday reality and endangered individual autonomy. As chapter 1 argues, thinkers writing for a broad though well-educated audience in the 1930s astutely identified the new culture of the twentieth century, radio's role in creating it, and the problems it posed. Broadcasting networks, they claimed, concentrated cultural authority in a few hands, and, in turn, disseminated uniform programs and messages to a mass public. The culture of the United States, these thinkers lamented, was shaped by fewer and fewer producers and increasingly designed to appeal to a common consumer: that is, it was increasingly mass-produced and intended for mass consumption. Such a mass culture, various thinkers argued, eroded personal distinction and centralized power and control of public communications. Taken as a whole, public intellectuals worried about the limits of individuals' abilities to shape their own lives and to speak publicly in the modern world.

Such critics were, to no small degree, correct. Unlike radio's contemporary defenders who saw broadcasting as entirely subject to the listener, these critics rightly noted that mass culture imposed constraints. But such critics told only part of the story. Cultural history has many shapers. In fact, Americans discovered limited space to invent radio's meanings for themselves. And in doing so they used the medium to help them address the very challenges mass culture fostered.

Radio did not drown listeners in a vast, impersonal quicksand. As chapters 2 and 3 argue, Americans found radio could help them personalize their encroaching mass society in order to find ways to count within it and to gain a measure of control in their own lives. Chapter 2 explores the ways listeners who tuned into popular programming forged personal relationships with radio's voices and found support within those relationships, making the intrusive wider world into something familiar. That process also made it possible for listeners to feel that they mattered in the distant but vital national political arena. To listeners, chapter 3 argues, political broadcasting, particularly Franklin Roosevelt's fireside chats, revived—and significantly revised—the idea of political participation. By reimagining the public voices they heard on the air in terms of private relationships, listeners found resources to help them negotiate their daily experiences and to restructure the democracy in which they lived.

When combined with radio's power to speak to a vast public, this sense of personal connection with distant speakers held out promise and posed risks, points chapter 4 assesses. The medium's intimacy and reach meant that some Americans found in radio the possibility of being heard in a massive public sphere. To some listeners, charismatic speakers with access to the nation's airwaves and with whom listeners felt a bond might serve as champions. But the line between popular champion and manipulative demagogue was not easy to draw. Radio's tendency to concentrate authority with such speakers might be—and would be—used both to stand up for ordinary individuals and give them a public voice, and to manipulate and further muffle common Americans. Both the potential of mass communication and the threats critics saw in it were real.

Despite the very real dangers within it, some still relished the promise of radio as mass communication because they understood communication with the masses as vital in modern America. Chapters 5 and 6 consider how some applauded broadcasting for making possible a new kind of communication, with a mass audience. By examining the academics who in the 1930s studied the medium, chapter 5 explores the possibility many of them saw that radio might allow at least a few speakers access to the public arena. In the process of essentially creating the field of media studies, these scholars implicitly debated the meanings of democracy and communications for the twentieth century. Chapter 6, the final chapter, looks at a group of writer-directors who came to believe that in the modern world, authentic communication had to reach a mass audience. To those writers who in the mid-1930s began experimenting with radio as a distinctly new art form, art mattered only as a public, not a private, expression. In the enormous modern world, these writers embraced the importance of communication that could speak to huge groups and, consequently, embraced radio. As many academics and artists saw it, radio might—might—offer some individuals otherwise-elusive meaningful public voices.

The story of those ambitious writers who took to the air also offers a reminder of the sharp limitations on the individual's ability to shape mass culture: in the end, corporate interests controlling the networks undercut writers' efforts to create a civic-minded public art. Clearly, as Americans looked to radio to help them navigate within the emerging modern world, they practiced a precarious balancing act. Radio, after all, was the very centralized, uniform, and vast culture many Americans struggled with in the Depression. We might, then, call some groups' understandings of radio at least contradictory: they turned to and relied upon one of the very sources of their dislocation. Moreover, from our vantage point, we can see that by using radio Americans did not overcome the challenges their mass culture posed. The seemingly private relationships listeners depended upon in their own lives were largely one-way and imagined; the voices some found on the air were at best strictly limited. Americans may have found ways to count in their world, but the meaning of democracy changed as public participation came to mean something more private and passive. Communication with the masses may have expanded access, but it shrank exchange. Ultimately, though, most found radio a valuable personal resource, not because they were dupes, not because they were blindly subservient to mass culture, but because they needed help to function within the world around them. Radio's America enmeshed individuals in an often-overwhelming world beyond their personal experience. Using radio, listeners had no room to tear down their unsettling mass culture; in point, radio reinforced it. But using radio, they found crucial ways to navigate within that world.

That would prove vital—in the 1930s and long afterward. In the decade's wake, Americans could hardly elude the tendrils of a mass society. The Depression was not simply an economic crisis, but a period of tremendous cultural change as well. The millions of Americans who incorporated radio into their lives in the 1930s struggled to find their places in the modern world. The rise of an extended society with increasingly centralizing and, in certain regards, standardizing tendencies was, of course, a process that began decades before the Depression and continued through the century. We should understand the 1930s as a part of that longer sweep. But in many respects, we can also see in this era a moment—and in radio, a vehicle—that drove home this new society. American communication and culture were forever changed. The experiences of those making sense of radio suggest that we should understand this as a period in which many Americans first truly inhabited and began coming to terms with the culture of the twentieth century. The daunting challenges that mass culture raised and the trade-offs listeners found in broadcasting would be staples of American life for generations. Radio and the Depression decade would make the emerging modern world virtually impossible to avoid.

The public intellectuals of the 1930s who observed the United States changing were not, in total, wrong: a culture that was mass-produced and designed for mass consumption—and hence in important regards centrally determined with common denominators in mind—did develop and reach the daily lives of many. That culture's arrival proved very unsettling. Even as many found their own meanings in radio, these thinkers were right that the changing culture posed real challenges for individuals seeking a measure of control in their own lives and, perhaps, a chance to count in their society more broadly. The emergence of a modern mass culture shook the bounds between public and private, the meaning of democracy, and the idea of communication.

The complex mass culture that public intellectuals identified in the Depression remained in the ascendant for decades to follow. Network broadcasting, for instance, embedded itself still more deeply into American life after the Depression, especially with the growth of television in the wake of World War II: what radio and the early twentieth century started, television and the mid-to-late century took to new heights. Radio in the 1930s did not represent the climax of the modern world, but, for many, an introduction to that world and to the problems it posed. The United States would spend much of the rest of the twentieth century within the legacy of its new mass culture.

Consequently, the precarious balancing act Americans practiced within that legacy became a vital part of many, many lives in the 1930s and beyond. The meanings of mass culture proved to be inflexible, but at the same time, not entirely absolute either. For all the easy rhetoric suggesting that popular choice drives a culture designed to appeal to the populace, those choices are starkly limited and do not include rejecting that culture. For all its influence, though, mass culture does not produce entirely predetermined and irresistible ideas, values, and messages. Diverse Americans in the 1930s responded to the structure of their radio system and the voices and stories they heard on the air by walking a fine line. Within very real constraints, they interpreted broadcasting for themselves. How they did so and what meanings they found in radio would be crucial questions.[17]

This process, of negotiating within a mass society to help shape one's own culture, became an essential one in the Depression. As people struggled to come to terms with the America radio did so much to create, they used that same mass medium to help them re-envision and navigate their mass culture. That effort to balance personal needs and a daunting public arena would become a central challenge not just in the Depression, but in the decades that followed. When journalist Dorothy Thompson reflected on Orson Welles's Halloween-eve presentation of *War of the Worlds*, she decided the program addressed the absolutely fundamental issue of her era. "The newspapers," she declared, "are correct in playing up this story over every other news event in the world." The Mercury Theater broadcast hinted at Americans' efforts in the 1930s to find a measure of autonomy in the emerging age. To Thompson, the whole episode was a poignant statement of the influence of a mass culture and the limits of individual control. She overstated the dominance of that mass culture, but not the enduring significance of the struggle.

"It is," Thompson wrote, "the story of the century."[18]

RADIO'S CHALLENGES

Public Intellectuals and the Problem of Mass Culture

From his vantage point squarely in late-nineteenth-century America, Edward Bellamy peered into the future and imagined a technological web that would allow people all over the country to hear the finest music and lectures in their own homes. Simply by touching a knob or two, Bellamy prophesied in his wildly popular novel *Looking Backward*, anyone anywhere would be able to listen to live performances any time of day. As the writer and social reformer depicted his hypothetical broadcasting system, it offered programs to suit the most distinctive tastes. The media of the future, he hinted, might make it possible for a range of voices to be heard. In 1888, Bellamy looked forward and delighted in the ideal of a means of communication that could span the country instantly.[1]

The reality proved far more ambiguous than Bellamy, social critic and aspiring prophet, ever imagined.

When modern broadcasting and modern America actually arrived in the 1930s, the newly real prospect of communicating instantaneously with millions took on a hue Bellamy could never have foreseen forty years earlier. The world had changed too dramatically. The twentieth century and the development of radio helped turn communication into a mass experience—one that shook the place of the individual in the United States. For all that radio opened up magical possibilities, it also transmitted modern mass culture's very real challenges. Radio might subsume the individual within modern mass society. It might deny individual distinction, speech, and autonomy, and instead speak to and for an artificial mass public.

Less than half a century after Bellamy had forecast that centralizing the production of culture could open up choice for individuals, America's public intellectuals looked unhappily at the arrival of a mass culture and rued the place of broadcasting within it. The modern world, as those thinkers who sought to reach a broad but well-educated audience astutely noted, was distinct because the twentieth century represented a moment when culture was both mass-produced and designed for mass consumption. More and more, these public intellectuals lamented in the 1930s, centralized authorities controlled and standardized ideas, values, and communication for the whole of the nation. And radio lay at the heart of this new culture, pumping it through the United States. Later in the twentieth century, the idea of a mass culture would become vaguer, encompassing many sorts of cultural forms. But in the 1930s, as public intellectuals critiqued the leading mass medium of their day, they offered a sharp sense of what was new about the emerging world, of what defined the mass culture they were coming to inhabit—and of the problems it posed.

These thinkers were right that the mass culture they identified would shape the century to come. They were right that radio contributed to the centralizing and homogenizing tendencies of modern society. And, perhaps most importantly, they were right that the new mass media posed real challenges: challenges for individuals seeking to find space for their own values and voices in the public arena; challenges for competing ideas of democracy. The medium helped push a vast, overwhelming sphere into people's lives; at the same time, ordinary Americans found it difficult to influence that world in meaningful ways.

As it turned out, though, public intellectuals were wrong about how ordinary Americans met those challenges and about how they used radio. Such critics never did sway or connect with the populace at large. But they were right that their era saw the intertwined rise of a mass media and a mass culture—and about the challenges facing individuals living with both in the twentieth century.

If they could make sense of radio, public intellectuals rightly understood, they could better comprehend the changes sweeping the twentieth century. As a whole, critics of the mass media precisely described—and assailed— mass culture in America. Mass culture was not simply culture for the masses or the introduction of certain technologies into the production of that culture, but a particular combination of the two: the easy reproduction and distribution of identical cultural forms, enabling a small slice of society to reach a vast and largely undifferentiated audience.

As these public intellectuals evaluated radio, they evaluated modern America. And these thinkers found plenty to worry about in both. Commentators with views on both sides of the political center forcefully expressed their disdain for America's fledgling broadcasting system. "In its use of the new means of communication, the land of opportunity looks more like the land of lost opportunities," the culturally conservative economist William Orton lamented caustically. On the left, Marxist poet and journalist James Rorty blasted radio for a hydra-headed assault on civilization itself. "Perhaps the scientific workers who developed and perfected the radio tube were . . . guileless as to motive," he asserted. "But in terms of social consequences, these playboys of the laboratories brought into the world hopes, apprehensions, marvels, and grotesqueries greater than they could have anticipated."[2]

Modern America, these thinkers maintained, was beset with the same problems. "The ether," Rorty liked to write, "is a mirror: this confusion of voices out of the air merely echoes our terrestrial confusion."[3] Such critics looked at the nation in the Depression and saw an enormous society interlaced with a web of invisible threads that distributed standardized products and concentrated authority.

As these public intellectuals might have explained it in their bleaker moments, mass culture fostered a vast, overwhelming, and intrusive landscape upon which only centralized authorities could leave meaningful marks. In defining the mass culture of their century, these thinkers clustered around and made visible two intertwined schools of thought: one concerned with the dangers posed by creating a single set of stories, ideas, values, and entertainments to be consumed by the overwhelming majority; the other, wary of centralizing the production of that culture in the hands of a few. Each raised vital problems for individuals in modern America.

In the former camp, critics such as William Orton attacked radio for fostering the mass consumption of culture. Radio's ability to reach the masses, these critics explained, led broadcasters to create only programs that appealed to an enormous, undifferentiated audience. Consequently, the mass media eroded elite cultural standards and the diversity of listeners; such thinkers feared that the homogenizing influences of mass culture menaced unique thought, creativity, and personal identity. As that mass culture, and the social changes it represented, intruded into everyday lives these critics worried it would melt individual excellence and distinction into a common sludge. If the only standards that counted were those of the populace as a whole, such thinkers wondered, were individuals no longer relevant? If so, what did democracy mean?

Others assailed radio from a more radical position, focusing on the impact a mass-produced culture had on social power in America. Commentators

such as James Rorty blamed the new broadcasting system for sharply concentrating authority. With access to the airwaves limited, only select groups
could effectively step out into a massive public sphere and make their voices
heard. For everyone else—the many who could not mass-produce their voices
so the entire nation could hear—free speech and meaningful political participation would become an anachronism, civic democracy a thing of the past. To
Rorty and most in this school, the major danger here was tied to capitalism:
as a national medium run by a few corporate interests, radio enhanced big
business's power to control society.

The risks inherent in a broadcasting system that narrowed control of the
production of culture are perhaps plainest in the thinking of several African
American public intellectuals. To these critics, it was painfully clear that the
centralization of radio networks meant blacks had limited means of shaping
the new national conversations, those that took place through the increasingly dominant medium. Many of these African American commentators
had experienced firsthand how a mass-produced media denied diverse
groups access to the ears of the public.

Of course, not all Americans, or even all intellectual commentators, worried about the new medium overpowering individuals. A notable minority
among public intellectuals praised radio—and the idea of mass culture in
general—as inherently empowering to audiences: the effort to attract a mass
audience, they claimed, meant popular tastes controlled broadcasting. Unlike radio's critics, the medium's defenders saw little to fear in the new medium, at least in part because they saw little to fear in a democracy focused
on masses rather than individuals. Democracy, they implicitly argued, did
not require personal diversity or individually empowered citizens, but majority rule. And the commercial system ensured this, such thinkers claimed:
listeners voted in blocs by turning their dials. The rule of the masses through
such a consumer democracy, these supporters of radio asserted, was truly
democratic.

Radio's intellectual defenders' idea that mass culture bowed to the will
of the populace had considerable endurance, but it ignored the realities of
broadcasting, modern America, and how power operated in a centralized
system. On the other side, those intellectuals critical of radio also overstated
the case, depicting an almost entirely oppressive mass culture. But their critiques—and implicit debate with radio's defenders—resonated nonetheless.
William Orton, James Rorty, and others accurately identified America's new
mass culture and recognized the genuine problems that culture posed for
individuals.

The two camps of intellectuals critical of radio, in fact, had much in common as they defined their mass culture: the world emerging by the 1930s

combined both mass production and mass consumption. Ultimately, most critics feared that mass culture was creating a mass public, at the expense of the individual and democracy. Radio, such commentators argued, helped give rise to an expanded public sphere composed of homogeneous aggregates instead of individuals. The mass medium pushed a new culture into people's daily lives, devaluing personal distinction and autonomy there; and it denied ordinary individuals a way to move beyond their personal circles and speak to the common polity. Meaningful pluralism of thought and meaningful free speech had no place in the mass culture radio's critics saw looming. As Americans integrated radio into their daily lives in the Depression, public intellectuals developed understandings of radio and its world that would influence critical approaches to broadcasting for decades. More than that, though, those understandings of the mass culture aptly described some of the challenges facing America in the 1930s and for the rest of the century.

William Orton and the Mass-Consumption Critique

For many of those public intellectuals critical of radio, the dangers they saw around them threatened to eliminate personal distinction and individuals' abilities to act meaningfully. The mass culture this group saw increasingly saturating their world was, at its heart, culture designed for mass consumption. The commercial network system prodded broadcasters to maximize their audiences, creating programs intended to capture the attention of huge blocs of listeners across the nation. In the pursuit of audience shares for advertisers, critics with this consumerist orientation observed, American radio distributed common stories and messages for a common taste. But such uniform thinking did not actually exist throughout the nation, they cried; radio invented it. Consequently, many of these thinkers lamented, the individual no longer counted: real people blurred into an abstract mass; distinction and action gave way to passive uniformity. This was no small fear. To commentators who often shared a classically liberal faith in unleashing the individual for the benefit of all, a program that threatened diversity of thought and the value of the individual would be a neutron bomb. The risks were real. What would happen to cultural and social progress in such a world? If personal efficacy and diversity eroded, did democracy wash away? Most fundamentally, if the public sphere existed in aggregate only, could one relate to the modern world as an individual? Those who analyzed a mass-consumed culture feared they knew the answer.

The future looked bleak to this group, in part, because the present sounded bleak. Those who staked out this critique tended to center their attacks upon the cultural form itself—the medium and particularly its programs—rather than focusing on either the broader, more systemic issues shaping the form or upon listeners. Critics of mass-consumed radio began by bemoaning the quality of what they could hear on the air. Most, however, moved beyond that, looking both to the causes and effects of that programming. They recognized that the commercial nature of broadcasting created an imperative to find and appeal to a mass audience. And it was this tendency of mass culture that doomed American broadcasting and endangered the place of the individual in modern society. Mass culture devalued personal distinction and free thought, mass-consumption critics said; and in doing so, it threatened to create a world with no room for democracy and cultural uplift. Instead of active thinkers and actors, radio called for a society made of passive consumers. To these commentators, this shift predated and extended far beyond network broadcasting, but radio had made the values and tendencies of a mass culture impossible to elude.

Of those critics, the most prolific and articulate on the subject of radio and culture for mass consumption was Smith College economist William Orton. Widely diverse thinkers, from the likes of Ring Lardner to historians Charles and Mary Beard, shared related views of radio, but none expressed the complexities of their common vision so thoroughly or cohesively in the 1930s. Born in England in 1889, Orton grew up immersed in high culture, giving public recitals on both piano and organ before he was twenty. After serving in World War I, he moved to the United States and Smith in 1922, where he taught and wrote until his death in 1952. In the 1930s Orton allied himself loosely with efforts by educators to reform radio: in addition to writing for a general audience, Orton occasionally addressed the moderate reform group, the National Advisory Council on Radio in Education. Never a political radical, Orton combined a concern for the individual with the acceptance of government planning to lay the foundation of his progressive, general-interest writings. According to Smith faculty remembering Orton when he died, his life and work was shaped by his belief in "the freedom of the human spirit in the true liberal tradition."[4]

Not surprisingly, then, in his evaluation of American radio, Orton revealed both his liberal convictions and a faith in the virtue of high culture. He was hardly alone. This strain of analysis drew upon both an evolving centrist liberal tradition and an elite conception of culture. Consequently, the consumption-centered critique of radio and mass culture tended to find voice in the day's more politically moderate and culturally conservative journals such as the *Atlantic Monthly* and *Harper's*. The 1930s produced a well-known flowering of writers and thinkers on the left, but for the most part, those intellectuals found expression in publications with more pronounced leftward political leanings, such as the *New Republic* and the *Nation*.[5] All of these journals—and most of their contributors—clustered around mainstream currents in American thought in the era. In a decade in which America tolerated radical left-leaning thought and the center's credo of liberalism evolved from its comparatively antistatist, classical meaning toward its more modern one welcoming government action, none of these journals championed conservative political stances. In terms of culture, however, the political liberals often took a more conservative view, regarding traditional high culture as superior to popular offerings. On the subject of radio these journals cannot be rigidly classified: writers readily expressed either point of view in either type of journal. In general, though, thinkers who leaned toward a mass-consumption critique of radio expressed mainstream liberal values rather than the more radical views of the Left in the 1930s. And, again in general, those commentators gravitated to traditional defenders of high culture.

For many in intellectual circles in the 1930s, their objections to radio began with an objection to the programs on the air: rarely have so many found the word "drivel" to be such an apt description. Writing in the leftist publication *Common Sense*, future historian Louis Filler found himself quoting journalist H. L. Mencken to express his frustration with the quality of radio broadcasts: "Here in America we get our radio entertainment for nothing; and that is exactly what it is worth." On the other side of a political fulcrum, the *New Yorker* assailed radio programs from its self-conscious position as chronicler and conservator of high culture. In 1932 and 1933, the *New Yorker* published commentaries on radio by the writer Ring Lardner. In his column, "Over the Waves," Lardner lashed out at popular programs and performers from *Amos 'n' Andy* to crooner Bing Crosby—who was, Lardner claimed, "extremely proficient at not hitting a note on the nose." Although Lardner admitted he enjoyed some programs, in general he believed radio programs represented a degeneracy of culture in America. Popular music on the radio particularly upset Lardner. "If ever there was a time in the country's history when song words needed a good missing, that time is (Bulova watch time) now," he wrote. Lardner saw himself trying to hold back a tide of new mores that had been rushing in over several decades. Radio was drowning in inferior music, Lardner felt; consequently, the airwaves threatened to swamp traditional standards of quality and decorum. "I don't like indecency in song or story," he wrote, "and sex appeal employed for financial gain in this manner makes me madder than anything except fruit salad."[6]

Despite the slap at fruit salad, Lardner's critique typically focused on the quality of radio's programs. Most public intellectuals, however, went beyond that sort of tony entertainment review, and considered the reasons for and implications of that programming. The trouble with radio, they argued, was that it was run by corporations for purely commercial interests. Programming was controlled not by trustees looking out for the public's interests, Orton explained, but "by persons concerned solely with making money out of the public." Orton did not take a conspiratorial view of this commercial oligopoly of the air. When confronted with the complexity and interconnection of twentieth-century America, Orton came to see a need for centralized regulation. The United States, though, had failed to plan, he said, and therefore provided the opportunity for commercial domination. "That this is so," he continued, "was due not so much to anybody's considered decision as to the lack of foresight and the slipshod inefficiency characteristic of the control of corporate life in the United States."[7]

Yoking radio to the pursuit of corporate profits, as the United Stated had, guaranteed listeners an inferior product, Orton maintained. Diverse commentators pointed out that commercial concerns demanded that broadcasters maximize their audiences. That meant, these critics asserted, creating unsophisticated programs that could appeal to even the lowest cultural tastes. Basing programming choices on popularity, then, did not denote a victory for a democratic culture—as defenders of mass culture claimed—but the loss of culture altogether. Music critic B. H. Haggin complained that radio stations avoided playing classical music because they feared exceeding the limits of the general public. In the name of increasing listenership, he wrote, radio producers made sure safe, simple musical forms dominated the air. The *New Republic* broadened this critique to suggest that all of radio programming was designed with general accessibility, not quality, in mind; and several years later the popularly known historians Charles and Mary Beard leveled the same charge. Even playwright Merrill Denison, who would later stand out as a public intellectual who actively defended radio, admitted in 1934 that the idea of maximizing audiences probably worked against the pursuit of high-quality programs. In the name of profits, mass culture critics and others agreed, broadcasters sought broadly popular programs—programs that were, consequently, inferior.[8]

But Orton and like-minded thinkers did not really consider lousy programs by themselves the underlying problem with this system of popularly dictated culture. Rather, at its very heart, their critique declared that disseminating only a single cultural source to be consumed by all segments of society fostered a uniformity of thought. To put this fundamental concern in other words, American radio might melt distinctive individuals into an impersonal collective. As advertisers planned programs that would maximize profits, they did not set out to appeal to low tastes; that was a side effect. Orton and others claimed that radio producers set out to appeal to a mass taste—that is, to smooth over differences between people's tastes and create a homogeneously approved product. Fully developed, then, this argument moved beyond a simple elitist defense of high culture; it attacked not popular culture per se, but a forced uniformity of culture and thought. Radio, these critics worried, treated a diverse collection of Americans as a single bloc. Multiple publics became a singular, the Public. Profit demanded radio appeal to the "mass-mind," Orton declared. But no such thing actually existed. The mass-mind, he wrote, was the creation of advertisers. "Society," Orton argued, "consists in reality of a very large number of distinct minorities, with different needs and different interests."[9]

Mass culture, Orton railed, attacked those distinct minorities; it ignored their different needs and different interests. The very concept of a mass mind devalued cultural and personal diversity. By implication, modern mass culture might be a social hurricane pounding the United States. Radio, Orton and others argued, neglected those individuals and groups who did not conform to a bland, standardized, and artificial common taste. By doing so, the medium excluded diverse people and groups from meaningful participation in the whole of American life. To Orton, suggesting that not all individuals counted in society threatened the very survival of a liberal democracy he conceived of as invaluable. A democratic society, as he saw it, prized real individual efficacy and freedom of thought; pluralism, not majority, should rule. "As a member of a not inconsiderable minority I [should] still [be able to] get enjoyment from my radio set for some hours everyday," Orton wrote. "And that is democracy. Where shall I find it in America?"[10]

Orton never indicated whether he meant to include racial or ethnic minorities among those squeezed out by a mass culture; African Americans, for example, certainly could have argued—and occasionally did—that radio programs overlooked their interests. He focused on what he considered cultural minorities, particularly those who valued high culture. But his reasoning could be more broadly applied. Mass culture did not conceive of people as individuals and thinkers, he said: only as undifferentiated consumers.[11] And when individuals had little standing in their world, the future of that world looked stormy.

The value of a democratic society, as Orton saw it, lay in its ability to free individuals to act as they chose. The classical liberal faith in free individuals as the source of human progress remained strong. To Orton and others, the opportunity to better the world was at stake. Many of the mass-consumption critics understood this risk first in cultural terms. As mass culture devalued the individual, they claimed, it also devalued individual creativity. Consequently, mass culture stifled the artistic and cultural progress Orton and others so revered. In a system in which popularity served as the measure of artistic success, creativity, excellence, and genius had no place, he lamented.

"To expect cultural leadership, artistic or intellectual pioneering, from the mass is more than even Mr. Coolidge would venture," Orton wrote. He believed in standards fixed in something firmer than the latest program ratings, and worried about the modern commercial world jackhammering that bedrock. Taking Orton's fears another direction, sociologist Jerome Davis suggested that mass and high culture could not coexist on the air. Bad culture, Davis wrote, would drive out the good because, in order to win listenership, mass culture teemed with excitement. By comparison, many found high culture dull, he observed.[12] The very presence of mass culture, debilitated attempts at cultural uplift.

To be sure, this was, on one level, a culturally conservative argument. These thinkers found modern mass culture too commercial, too banal, too fast-paced: all that the rarified world of culture was not supposed to be. Charles and Mary Beard shuddered at the sounds that streamed from their radio sets—and at what those programs drowned out. "Now the canned rumbles, thumps, and rattles poured out of radio sets, unremittingly and ceaselessly," they wrote. "Amid all the din, however, one thing could not be refuted: contemplation, meditation, and quiet reading were becoming increasingly difficult." The intrusion of commercialism and popular tastes into the cultural and artistic world offended these critics' sense of order. They clung to a Victorian notion of separate spheres, and despised radio for violating the divide between humanity's base economic concerns and lofty cultural ones. "The association of cultural programs—such as opera broadcasts—with commercial salesmanship is inherently degrading to art and artists, and is likely to do harm rather than good in the long run," Orton thundered. "In my view, no compromise is possible on this question."[13] Again, this was, on one level, a culturally conservative argument. But that was not all it was.

Despite Orton's declamation, this was never simply a struggle for better art. Orton placed cultural progress near the center of the human mission. As mass culture threatened that progress by demanding conformity, he suggested, it demeaned humanity. Orton and others believed in the Victorian idea of art and high culture as certain and elevating. He disdained radio in the United States, in part, because it belonged to a world that rejected such thinking. And this, he believed, endangered the value of human existence. The supremacy of the profit motive, Orton wrote, "renders social life progressively more meaningless and more brutal." To varying degrees, others, including Rorty, shared Orton's fear for the future of life in America in the face of mass culture's shifting popular standards. "When this idea of 'let the people rule' is uncritically applied in education, what happens is that first education perishes and eventually civilization perishes," Rorty wrote. It was individuals who pushed a society forward, intellectually and socially as well as artistically. If America did not allow for individuals to stand out, to make their own way, to push their culture—Orton and others doubted if society could survive. "The redemption of the mass," Orton wrote, "cannot come except from minorities."[14]

As Rorty implied, the social risks inherent in basing cultural standards on that which attracts the largest body of consumers were most visible in the arena of education. As mass-consumption critics discounted radio's much-trumpeted educational potential for precisely this reason, they illustrated their real concern that culture for mass consumption undermined social progress. As long as radio sought to educate the masses, Orton and other commentators claimed, educators would have to water down their content and spice up their presentation to such a degree that a program could offer little of value. The desire to teach the mass mind meant that radio would not offer anything beyond the grasp of a thirteen-year-old, Orton complained. Moreover, since listeners could easily tune out, educators would have to sell their programs, just like commercial ones. That meant eliminating rigor and challenging or complex ideas. It meant condensing education into sensational entertainment. It meant, conservative editor Travis Hoke wrote, "The new cultural process will be pleasant and tedium will be gone. . . . For it will be discovered that the 'radiot' . . . will not listen long nor to big words, and cannot be forced to stay in class nor after school. It is too easy to flip the dial to another station. . . . Doubtless the day will soon be at hand when five minutes will be enough for Einstein, theme song and all."

In other words, critics noted, it meant dissolving the benefits of education in a mass-produced syrup. Students and audiences would evaluate education based on its palatability, not intellectual value. Such an indictment had power: half a century later another public intellectual, media theorist Neil Postman, would level essentially the same accusation against the leading mass media form of his day, television. The fact is, Travis Hoke charged in 1937, "a thing can go by the name of education and still be worthless."[15]

To question radio's value as an educational device in the 1930s amounted to a serious attack on the medium. Radio executives, politicians, and educators all raved about radio's educational promise. Even many of those who wanted to reform the existing system believed that radio, in the ideal, could revolutionize learning, bringing at least a base level of education to isolated reaches of the country. This is no doubt why Orton and other mass-consumption critics took care to explain the general dangers of culture designed for mass consumption in terms of education. Radio, they acknowledged, disseminated information to people flung across the nation. But to what end? "Where do educators get the idea that radio is a promising educational medium?" Hoke blasted.[16]

The very form of radio—not simply the commercial system as some less critical commentators claimed, but the nature of the medium as one for mass consumption—made it incompatible with education. Radio, Orton charged, encouraged passivity: it fostered the acceptance of standardized ideas and failed to inspire free analytical thought. Using radio to educate, he implied, meant educating listeners to absorb homogeneous ideas produced for all, rather than to develop diverse ideas on their own. "Radio has brought no new asset of major importance to education, and its use involves serious disabilities," he wrote. Since the first scheduled radio broadcast in 1920, the editors of the New Republic noted, they had heard all about radio's educational promise. The medium had not delivered, they declared: "The present trend of educational theory is away from learning by mere sitting back and listening—all that radio has thus far been able to provide." Radio might be able to disseminate information, but such distance learning could not demand listeners discipline their minds or think on their own.[17]

More than education was at stake here; the critique of radio as teacher vividly illustrated Orton's larger point. The medium, he worried, generally promoted a passive populace. Mass culture declared that ordinary people were consumers, not creators, of culture. A mass medium pumping entertainment into the home at all hours, for instance, enabled Americans to spend leisure time listening instead of doing, Orton noted. Music fans, he suggested, no longer took an active role in shaping the melodies they heard. "Any summer evening of the 1920's, suburban streets were enlivened by Millie's efforts to render the new 'song hit' on the family piano," Orton wrote in 1936. "The modern Millie flops down by the radio."[18] A homogeneous culture for all would not require independent action or thought. There would be no need—or space—to pick one's own path in the world.

Nearly fifty years earlier, when Edward Bellamy had imagined a broadcasting system, the utopian writer cheered standardized musical broadcasting in part because it meant the Millies of the world would not have to sing or play. Anyone could hear well-played music anytime.[19] From Orton's viewpoint, anything that discouraged personal differentiation and creativity seemed ominous at best. The economist worried about radio blurring passive individuals into a standard mass. This was a twentieth-century concern, one Bellamy's 1880s did not quite raise.

To Orton, the modern world justified this fear. He looked around and saw decades of change gnawing away at the individualism that he so valued in the United States. Americans' experiences, he believed, were taking place in larger and larger spheres, meaning standardized and abstract interactions came to replace personal ones. Traditionally, Orton explained, the genius of the United States lay in its meaningful face-to-face interactions. "American individualism," he wrote in praise, "means, historically and dynamically, an instinctive preference for the concrete, personal, spontaneous process of community life over the abstract, general, artificial processes of law, politics, and, above all, finance." Anything that threatened this individualism—and the many small, personal publics that fostered it—consequently threatened the heart of America. More and more, he worried, remote, impersonal, and homogeneous forces intruded on Americans' lives. More and more, Americans felt the anonymous grip of a central government, a national economy, a New York–based culture. The United States, Orton still maintained in the early 1930s, consisted of many local communities linked to a central whole by thin wires. But those links were becoming more intrusive; an artificial and standardized life was taking hold.[20]

That new world, these commentators lamented, was inescapable. Mass culture intruded into the heart of daily life. Radio transformed the air itself into a subversive agent of unwelcome change; it could infiltrate even the home and all private space. "The wholesale exploitation of sound in the various perversions of money getting is a far worse thing than the desecration of the countryside by billboards," Orton wrote. "It is at once more intimate and more degrading." The endless commercial cacophony, some worried, might transmute humanity. And even turning off your set might not solve the problem; radio allowed one nowhere to hide from the new mass culture. "Whether you hear them or not," wailed novelist Irving Fineman, "those incessant programs penetrate your flesh and blood, you breathe them in."[21]

You breathe them in—and once you do so, Orton and others feared, you might virtually cease to count as an individual in your society. A homogeneous culture designed for a homogeneous populace was inescapable, they suggested, and it hallmarked modern life. Broadcasting treated separate individuals and groups as part of an undifferentiated collective. The mass media demanded passive conformity and rejected distinction, and, in turn, the values of liberal democracy and progress. Culture for mass consumption was only half of the mass culture framework public intellectuals constructed as radio swept the United States, but the critique cut sharply in the Depression. And it would continue to hold weight for years to come. In the wake of World War II, intellectuals would level similar charges. In his influential postwar analysis, for instance, critic Dwight Macdonald revoiced Orton's fears about mass culture dissolving the individual. And, although less immediately connected, the wave of cold war–era thinkers who rediscovered Alexis de Tocqueville and the idea of tyranny of the majority, and who wrote about suburban conformity, expressed related concerns about the homogenizing influence of modern life.[22]

It was, and would be, easy, of course, for opponents of Orton's thinking to dismiss such ideas as cultural elitism. But as Orton articulated it, the critique of culture for mass consumption was never only an attack on the quality of programming. He and others who shared this view explored the standardizing impulses that they found in radio in order to consider the very real challenges the twentieth century posed. To those critics of a medium made for mass consumption, it seemed an enormous and abstract world intruded dangerously upon individuals, eroding free thought and the possibility of taking meaningful actions in their own lives. Not all public intellectuals critical of radio in the 1930s agreed on the precise nature of the era's challenges, though. For another group of such thinkers, the question of what audiences heard on the air paled before the matter of who controlled it.

James Rorty and the Mass-Production Critique

The mass consumption of radio expressed only one side of the interlocking story of mass culture. To a swath of often more radical public intellectuals, dwelling too exclusively on radio's creation of a mass audience obscured what they saw as the larger issue at hand: individuals losing the ability to reach out and speak in the public world. To these critics, mass culture was, at its core, culture that had been mass-produced. Technology made it possible for a very few producers to create and disseminate cultural forms to the whole of the country. And technology coupled with the social configuration of the United States made such centralized control of culture not just a possibility, but, increasingly, the reality. To an array of commentators—including both thinkers on the political left and some African American critics—that posed serious problems. As culture was increasingly mass-produced, only a select group had access to the technology that could enable them to disseminate their ideas, values, and stories to an emerging national public. Most Americans were left out of the process of creating culture; as public conversations became national ones, most Americans were left without a voice. Real freedom of speech was at stake. And if ordinary people could not speak out and be heard publicly, critics of a mass-produced culture suggested, democracy was at risk.

In the 1930s most public intellectuals understood this threat in economic terms: the forces of big business dominated radio and denied individuals or opposing views a meaningful voice. To an array of African American commentators, the danger was racially driven: the narrow control of the airwaves did not allow blacks to join in public discussion. These two groups of critics did not speak directly to one another on this issue; they saw very different particular forces undermining democracy in the United States. But on a fundamental level, both agreed on and were right about the underlying risk: the centralized control of culture they saw in radio threatened to deny individuals voices in the modern world.

Unlike Orton and those who focused on the mass consumption of culture, those critics who understood mass culture in terms of production did not automatically dismiss the idea of a medium that communicated with the masses. The important matter to consider, for the mass-production critics, was who controlled that means of communication. Instead of dwelling on the homogenization of culture and the audience, these thinkers focused foremost on the centralization of power and speech. As most of these commentators saw it, radio had concentrated the control of culture in the hands of an extremely undemocratic minority: the captains of capitalism. As businesses mass-produced not only goods, but ideas, these critics lamented, corporations tightened their influence over thought in America. Corporate control of radio made a mockery of freedom of speech; only big business had access to the public ears and airwaves. To many of these leftist critics, when the many voices of the people gave way to the voices of a few propagandists, this overtly signaled the erosion of democracy. This was a critique at once more and less forgiving than Orton's. Mass production's critics saw some hope for radio; they had faith in reaching a mass audience. But at the same time, this group had much less tolerance for the American social and economic system. In the United States, they suggested, radio's flaws replicated and exacerbated inequalities that pervaded capitalism itself.[23] Mass culture, their critique of radio and capitalism implicitly suggested, could not be separated from the balance of power and voices that produced it.

The Depression pushed leftist intellectuals to vocalize their critiques of the United States more actively and made it possible for them to do so. Amidst the dire American economic crisis, writers in influential opinion journals like the *New Republic* and the *Nation* could call for left-wing and Marxist-leaning reforms without slipping out of generally acceptable intellectual currents. Such journals provided space for what were, at least in the 1930s, vaguely mainstream challenges to capitalism.[24] A diverse array of intellectuals considered radio from this political ground. None, however, evaluated it more thoroughly than journalist and poet James Rorty. Only author Ruth Brindze came close to explaining the capitalist critique of the medium as fully and to spreading that critique as widely as Rorty. Born in New York one year after Orton in 1890, Rorty, like Orton, engaged in graduate studies and served in World War I. After the war, Rorty worked intermittently as an advertising copy writer and as a journalist, including helping to found the *New Masses*. By the Depression he had abandoned advertising, but his knowledge of the field would inform his critiques of American society and radio through the

1930s. Again like Orton, Rorty affiliated loosely with educators' push for radio reform; Rorty, however, sympathized with the most radical of the reform organizations, the National Committee on Education by Radio. Obviously, Rorty's views reflected his political stand. Although Rorty shifted his party identification from the Communists to the Socialists over the course of the 1930s, he remained a committed Marxist critic of capitalism—and of its influence over broadcasting.[25]

Rorty, Brindze, and others in this school had little praise for programs on the air, but unlike the consumerist-minded critics, they saw no reason to start their critiques there. The real issue, believed Rorty and those who shared his thinking, was not what was on the air, but who controlled the air. After all, radio dramatically concentrated that control. As a mass medium, radio had the power to influence listeners throughout the country—and therein lay its promise and its pitfall. From one broadcasting microphone, a single speaker could reach the minds of millions. The technological and systemic realities behind the mass production and distribution of sound through network broadcasting helped define mass culture—and made the question of who could use the mass medium to amplify their voices a crucial one. In modern America, radio was, Rorty claimed, "our major instrument of social communication." And, he suggested, with decades of social and economic upheaval and the Depression disrupting the social and political order of the United States, "social communication" had a vital role to play in reshaping that order. By the mid-1930s, radio was, Brindze declared, "American's no. 1 instrument for control of the public mind." Jerome Davis, expressing the leftist intellectuals' sense of radio's potential as well as threat, put the point equally bluntly: "Whoever owns the agencies for the distribution of ideas is most likely to control the people. Radio today ranks as perhaps the most important force for the dissemination of ideas in American life." And this, he explained, posed a problem. "We have," Davis noted, "permitted this incalculably valuable and powerful tool to fall into the hands of the power-trust group, which includes the radio trust."[26]

For left-leaning public intellectuals, then, radio's problems were structural. The broadcasting industry had fallen into the hands of businesses producing programs purely for private gain. From almost the earliest broadcasts on, Rorty wrote, businesses had enslaved the medium. "The whole art of radio was originally conceived of as a sales device," he explained. Ruth Brindze echoed that viewpoint: "Broadcasting in America has always been an industry whose primary purpose has not been public service but private profit." The commercial masters of the radio industry, Rorty allowed, secured their hold subtly, encouraging listeners to think those who twirled the dials had the final say over radio. Rorty and others, however, dismissed the idea that by tuning out or by writing a letter to a station, a listener could control the airwaves. Baloney, Rorty said; commercial interests would always win out over a listener's desires if the two clashed. Radio was no different than any of the technological advances that had engulfed America in recent decades. "Every genie, such as radio, that pops out of the laboratory bottle of modern science," he lamented, "is [put] to work making money for whoever happens to hold the neck of the bottle."[27]

The trouble, from the point of view of these radical thinkers, was not simply that radio operated on a commercial basis, but also that only a few corporations controlled the industry. The structure of broadcasting in the United States did not allow for competing stations or voices; the two commercial networks held a near-absolute monarchy of the air. Even though only about a third of the nation's stations affiliated with NBC and CBS, these critics noted that the two networks ruled almost 90 percent of the nation's transmitting power because they controlled most of the high-power stations across the country. To writers such as Rorty and Brindze, this monopoly of the airwaves guaranteed that radio would serve the interests of corporate capitalism in America. Beyond appeasing business because of their reliance on advertising revenues, the networks were big business in their own rights and were owned by the same huge financial powers that controlled many of America's banks, power trusts, and corporations, the crusading writers noted. Network domination of the ether left almost no room for any broadcasters who did not share the commercial cause, these critics charged.[28]

The structure of American broadcasting was particularly invidious because it essentially enabled a few corporations to control the most popular sources of political information and entertainment throughout the country. Unlike other forms of mass production in which a few industry leaders controlled goods, the centralization of radio production gave a commercial elite control over a critical segment of American communication—and thus, mass production's critics stressed, control over American thought. In the American broadcasting system, they argued, a few central authorities had the power to censor and dictate on-air discussion, and, in turn, to shape listeners' values and ideas. Those who controlled the networks, these thinkers lamented, decreed what subjects broadcasters could address and what points of view programs could support. In playwright George Kaufman's light short story, "God Gets an Idea," the Almighty himself is censored by radio's powers that be. Wanting to do something about the problems on Earth, God decides to go on the air to talk to Americans. The plan runs aground when a radio advertising executive refuses to allow God to discuss controversial subjects like religion and evil on the radio.[29] When programs that could reach the whole of the country could be mass-produced under the direction of a few authorities, only that select few could set the terms for discussion.

Because of radio's powerful national reach, commentators on the left believed that allowing one group sway over radio's content in that manner gave that group a staggering influence over American tastes and values. "The radio," the *New Republic* editorialized, "is an instrument, not for the free formulation of public opinion, but for molding it to suit the purposes of the small group of men who control the most important aspects of our national economic life." America seemed in the midst of trading away the very concept of democratic free thought, these critics feared. "Do you realize Ladies and Gentlemen of the Great Radio Audience," Rorty wrote, "that your ears and your minds are offered for sale to the highest bidder . . . ?"[30]

That highest bidder then used the medium as its own propaganda machine. Rorty, Brindze, and others looked at broadcasting in the United States and saw a system in which a corporate America centralized its control of the airwaves and used that power to mass-produce and disseminate culture to imbue their society with the values of consumer capitalism. The Depression could have led Americans to question their political and economic system, but radio gave capitalism an additional influential means of promoting its own ideals. Such consumerist propaganda secured the dominance of commercial values in the twentieth century, Rorty wrote: "Advertising has to do with the shaping of economic, social, moral and ethical patterns of the community into serviceable conformity with the profit-making interests of advertisers and of the advertising business." And by the 1930s, Rorty explained, radio had become one of business and finance's major propaganda instruments. For decades, critics on the left had worried about business's influence, of course, but radio had enabled corporate America to disseminate its ideology still more effectively. With America's consumer system at risk in the Depression, commercial leaders more and more enlisted radio in their effort to convince consumers of the virtues of that system. The flood of ads on the radio, for instance, taught listeners that they could buy solutions to their problems, writers such as Brindze and the *New Republic*'s T. R. Carskadon charged. Even a nonradical like liberal William Orton recognized the corporate control of American values and radio's significance as a propaganda source supporting that control. "Big business has in fact come to occupy in America very much the position occupied by the Church in mediaeval Europe," he wrote. "[It] moulds the forms and sets the standards of social intercourse, permeates while it patronizes the national culture in a hundred ways."[31]

Orton, of course, primarily feared that giving business such influence would degrade individual distinction. Those who criticized radio from the point of view of mass-producing culture worried far more about mass culture's potential to crush meaningful free speech and, in turn, democracy as a civic practice. The monotone on the air would drown out diverse individual voices. Since access to the powerful propaganda machine of the airwaves depended on one's bank account, Rorty explained, radio further helped transform financial power into social and political power. Freedom of speech in the United States, Brindze pointed out, did not include free access to the airwaves. In an era in which the most important public discussions took place on those airwaves—as these critics believed—if ordinary citizens could not get their views voiced on the air, their freedom to participate actively in civic conversations was, at best, limited.[32] As long as select business powers controlled the day's leading means of public expression, most Americans would find the value of freedom of speech replaced by the freedom to listen.

This was no small matter. Freedom of speech, Brindze declared, lay at the foundation of democracy: without open opportunities to influence public opinion, the voice and votes of the people could express nothing more than a squawk of a parrot. Theirs was a civic vision of democracy: a venue in which all citizens participate in public debate, actively discussing public matters, not simply voting upon them. "Freedom of speech is a correlative of democracy," Brindze argued. "It is remarkable then that its perpetuation should be entrusted to a few monopolists." But thanks to radio, she and Rorty asserted, that was exactly what had happened. And democracy suffered.[33]

Any system of mass-produced culture endangered democracy by narrowing participation in public discussion, such critics suggested. But America's posed special problems, these public intellectuals charged, because of the particular voices radio amplified. Would a radio system that bestowed tremendous influence upon big business "reenforce economic conservatism, strengthen vulgarity, and drive the American mind to an undemocratic Right?" Charles and Mary Beard asked rhetorically. To some of these thinkers, radio appeared undemocratic, in part, because corporate power and the interests of big business appeared undemocratic. Their broadcasting system inspired fears of European-style fascism in the United States. The hallmark of fascism—at home or in Germany and Italy—many suggested in the mid-1930s, was not governmental dictatorship, but an undemocratic rule on behalf of the interests of economic powers. World War II forced these commentators to reconsider that stand later, but during much of the 1930s they saw unchecked capitalism as the leading forerunner of fascism and a threat to democracy. Consequently, corporate control of radio alarmed those who shared the twin leftist values of the 1930s of economic equity and international antifascism. In European nations the government censored the air, the *New Republic* reported. But, the editors asked, was that really worse than censorship by "the ultra-conservative public utility magnates who for the most part govern our airwaves?" Because of the technology involved, the control of radio would inevitably centralize, Louis Filler asserted in *Common Sense*. The only question that remained in his mind was whether or not that centralized control would emerge as a private monopoly with the potential for fascism. To these thinkers, modern corporate capitalism was antithetical to liberty and democracy; they equated a worldwide rise in fascism with the modern expansion of big business's social control—a control American radio readily facilitated. "Broadcasting is controlled by our moguls of business and finance. This is the class which in Italy and Germany has benefited most from that new form of government known as fascism," Brindze wrote. "If fascism ever happens here, the new leaders will not have to seize the radio; they already control it."[34]

Rorty, Brindze, and others were clearly motivated by their political and economic ideology in this attack on corporate radio. But theirs was also a point of view that included a structural critique of broadcasting and elaborated upon Rorty's vision of democracy in a mass society. Democracy, in this view, was an active process, in which diverse citizens could voice their own agendas. The idea that American might exercise their authority by choosing among mass-produced products was an illusion, Rorty implied. A vision in which listeners controlled the air by voting with their tuning dials wrongly equated consumer choice with democratic authority. In setting up American broadcasting, though, federal regulators accepted that illusion as fact and consequently touted an undemocratic consumer version of mass democracy. Federal law required that radio serve the public interest. But, Rorty bemoaned, regulators and broadcasters wrongly interpreted the public interest clause to mean that which interested the public. Popular programs, he wrote, did not automatically serve the public interest; popularity was easily manipulated by the corporate authorities who controlled the air. Other, more civic-oriented considerations needed to come into play. Rorty might or might not have questioned whether there was even a singular public whose interests could be neatly identified, but he had little doubt whose interests were served by measuring radio with a popularity standard. Rorty—like Orton, though on quite different grounds—bluntly rejected the idea that the pursuit of popular programming fostered a democratic medium.[35]

That is not to say that Rorty bluntly rejected the idea of radio in any capacity. Those who critiqued a mass-produced culture focused on the structure that produced the culture; change that structure, and radio might serve positive ends. In the ideal, a medium that could communicate powerfully with millions of listeners was neither good nor bad. In this regard, of course, production-minded critics like Rorty differed sharply from consumption-focused ones like Orton. Both groups disdained the mass culture of modern America, but each understood the essence of that new culture differently and only together did they fully define it. To Rorty and his fellows, the primary problem with mass media was not its tendency to treat people as an undifferentiated bloc, but the centralized control it facilitated. Consequently, where Orton condemned radio as communication for the many, Rorty and like-minded thinkers applauded the unfulfilled potential of a medium that could reach the multitudes.

This discrepancy and the mass-production critics' enthusiasm for the possibility of communication that included the bulk of ordinary Americans were particularly visible in the thinkers' ideas of education on the air. Where Orton generally dismissed radio as an educational medium, intellectuals on the left occasionally trumpeted broadcasting's impressive—if almost entirely untapped—educational potential. Bruce Bliven, editor of the *New Republic*, jabbed at American radio for doing nothing to educate listeners. Like many intellectuals, though, he pointed to Britain and the BBC to show that in the right hands, radio might serve the cause of popular learning. And Jerome Davis claimed that radio was absolutely essential to the critical task of adult education in the United States. None of these radical critics believed American radio actually fulfilled their educational agendas in the slightest, but the potential for popular education potential they saw in the medium excited them. Rorty wrote, "The radio looks to me like the most revolutionary instrument of communication ever placed in human hands; it seems to me that its free and creative use, not to make money, but to further education and culture and to inform public opinion is perhaps the most crucial problem with which our civilization is confronted."[36]

Rorty believed in radio's educational potential. He had his doubts, though, about whether education itself was enough to save America. Could education as it currently existed do anything more than teach people to fit neatly into that undemocratic, dehumanizing capitalist culture? he wondered. In his bleaker moments, Rorty suggested that giving educators access to the air would do little to reform society: educators too had caught values of capitalism, and would only help train people to work within those values.[37]

Since the late nineteenth century, Rorty believed, scientific and economic changes had completely reconfigured America, making possible new levels of centralized authority. In the 1880s, Bellamy had imagined that consolidating production might give ordinary people a greater public voice; in the 1930s, to Rorty it appeared that centralization had instead given select businesses never-before-seen influence. To Rorty, the mass production of culture embodied the widespread dominance of corporate capitalism in the modern United States. As he saw it, one basic question faced the whole of American

life in the twentieth century: should America's resources profit private ex-
ploiters or serve the public interest?[38] Rorty feared that as Americans tried
to answer that question, outmoded laissez-faire thinking made it difficult for
many to grasp the distinction between the two.

In his view, America's ideas and social tactics no longer matched the
nation's reality. Only by overcoming that cultural lag, he explained, could
America be saved. Radio theoretically might offer an answer, but as config-
ured in the United States, it only exacerbated the problem. "Granted that
radio is socially and politically one of the most revolutionary additions to the
pool of human resources in all history," he wrote: "how does one go about integrating it with a civilization which itself functions with increasing difficulty
and precariousness?" Rorty looked at the American broadcasting system
and saw mass production with centralized commercial authority, modern
capitalism, spinning out of control; he looked at America and saw the same
thing. "It may be," he claimed, "that at bottom this chaos is merely a phase
of the conflict between science and politics, between industry and business,
between ownership and management, between class and class, between our
advanced technological means and our obsolete social and economic mores
and institutions."[39]

Radio, Rorty frequently declared, mirrored America. These critics of radio
found American broadcasting threatening and corrupt because it replicated
and expanded a social order that they found threatening and corrupt. They
disdained radio because they disdained the centralized corporate society
they saw gripping America. Ultimately they wondered how individuals could
be heard in a public life in which a few authorities controlled the discussion.
Like the consumption-focused thinkers, the mass-production critics tended
to leave ordinary listeners out of their assessments of mass culture. They
would have explained their emphasis on the broadcasting industry's structure by pointing to its power: listeners had little room to resist radio's messages, they believed. These commentators may have overstated the point, but
it was an evaluation that raised compelling questions, questions some critics
on the left would continue to ask for decades. Followers of the Frankfurt
School, for instance, would give voice to some pieces of this critique into
the 1960s. This was, however, an evaluation that made most Americans, intellectual or otherwise, increasingly uncomfortable as the political climate
changed after the Depression. Rorty, Brindze, and others challenged some
of the basic tenants organizing American society and capitalism. In the face
of World War II and, later, the anti-Marxist climate of the cold war, they
would find that radical analysis more and more difficult to maintain.

The critics of a mass-produced culture may have seen their thinking
downplayed as their economic ideology lost step with many in the United

States, but the cultural concerns they raised in the 1930s were, indeed, real. A centralized mass culture posed a genuine threat to individuals' voices in that culture. Americans seeking to speak publicly in the twentieth century had to confront that problem. The experiences and evaluations of African American public intellectuals reflecting on this system made that plain.

African American Intellectuals and the Mass-Production Critique in Action

African American thinkers who criticized radio in the 1930s would have, in a passing conversation, found little common ground with Rorty, Brindze, and their fellows. To many black intellectuals who commented publicly on radio, race, not corporate capitalism, was the crucial matter for discussion. But as select African Americans assailed radio for its racial politics, they too articulated the risky implications of a mass-produced culture that made possible centralized control. They found they had very limited access to or influence over white-controlled network airwaves, and that presented a challenge for finding a voice in an increasingly national public arena.

For the most part, African American intellectuals did not dwell on the issue of radio. Over the entire Depression decade, the *Crisis*, the NAACP's journal, ran just over half a dozen articles that related to radio, and the National Urban League's journal, *Opportunity*, ran none.[40] But occasionally individuals addressed radio; at those points, they typically confronted the medium's racial failings—and the lack of opportunities for African Americans to correct those failings on the air. Under the leadership of its editor and publisher, Robert Vann, the *Pittsburgh Courier* blasted radio, particularly the stunningly popular *Amos 'n' Andy*, for portraying only negative stereotypes of blacks. Vann attempted to push broadcasting to change that portrayal, but his cause produced little concrete effect. To the poet Langston Hughes this was precisely why American broadcasting was among the least free forums of speech in the world. Hughes had also tried to gain a voice on the national airwaves and found himself shut out. Like Rorty and Brindze, such critics did not object to radio in the abstract; they saw it as a highly influential medium. In practice, though, as long as only a few authorities controlled the air, these commentators argued, blacks could be, and often were, excluded from a national conversation.

During the early 1930s, as the *Pittsburgh Courier* mounted a campaign to chase *Amos 'n' Andy* from the air, black intellectuals brought together under the *Courier*'s umbrella critiqued the narrow control of culture that radio fostered. Network broadcasting, voices in the *Courier* charged, had an exceptional power to mass-produce and to spread images. And only the small group who controlled that process had the power to determine what those images would be. To the *Courier*'s commentators, radio became a national venue for racist propaganda, one that allowed no room for dissenting views. Amos and Andy were among the very few African American characters that network executives allowed on the air; because blacks could not challenge those portrayals, *Courier* writers warned, America came to see all African Americans in light of the program's portrayal. The images *Amos 'n' Andy* spread were not new, but the means were: the mass medium extended the reach of old racial stereotypes and presented those images in a popular venue largely closed to blacks.

In an attempt to remedy these problems—both to curb the racist imagery and, implicitly, to gain a sort of voice in the airwaves' public conversation—in 1931, the *Courier* turned its editorial stance into an active drive, collecting over a half a million signatures on an anti–*Amos 'n' Andy* petition. Certainly, the *Courier*'s stand represented the popular newspaper's effort to sway mass opinion, not just to develop an intellectual critique of radio. But the actors behind the campaign were some of the elite commentators in African American journalism in the 1930s and they sought not just a popular reaction, but an analysis of radio in action. More than a year before the *Courier*'s petition drive, critics in the paper attacked radio for spreading damaging ideas about African Americans through *Amos 'n' Andy*. The popular program made blacks seem different and inferior, giving segregationists a boost, said Bishop William Walls in a speech liberally quoted in the *Courier*. "The movement of segregation and aversion to the black man is shrewdly and sinisterly carried on by quiet maneuvers in politics, economics, publicity, the stage, the press and now the radio," said Walls, head of education for the AME Zion Church. The *Courier*'s lead columnist, George Schuyler, agreed, complaining that the radio program simply affirmed whites in their stereotypes of blacks.[41]

The problem was not simply that the program portrayed blacks negatively. The problem was that blacks also had no means to offer alternative images to the national audience. Because of networks' prejudice and their absolute control of national radio, African Americans could not get on the air to counter the negative images in *Amos 'n' Andy*, suggested Floyd Calvin, who later replaced Schuyler on the *Courier* editorial page. "Color prejudice prevents Negroes of talent from giving to the same radio audience a corresponding picture of the better side of Negro life," he wrote. Consequently, white America never heard about blacks except as ridiculous clowns. If blacks could speak freely on the air, Calvin said, *Amos 'n' Andy* would still deserve criticism, but it would be legitimate speech; in that case, blacks would have the opportunity to counter stereotypes by speaking out and providing their own racial images and context. Without that, though,

Courier editorials periodically noted during the petition drive, whites often uncritically believed "the machine of dirty propaganda."[42]

As the *Courier*'s active campaign against *Amos 'n' Andy* wound down, Vann subtly shifted the emphasis to focus on boosting self-respect among African Americans. Perhaps he realized the unlikelihood of toppling the most popular program in America, and revised his primary goals. As Vann changed his campaign's ends, he implicitly accepted the central control of the mass medium and the ways that limited participation in mass communication. Instead, Vann essentially sought to take advantage of an alternative discourse, one with a narrower audience but that offered freer and more meaningful speech for him. By the fall of 1931, the *Courier* took to referring to its fight against *Amos 'n' Andy* not in terms of programming, but as a crusade for self-respect. The paper less often couched its drive in terms of changing what all of America heard, and more often referred to changing how African Americans thought about themselves.[43] This new emphasis had precedent, however. Commentators in the *Courier* had worried for several years that listening to popular radio programs might implicitly cause some blacks to accept the disparaging portrayals. In the summer of 1930 Schuyler scolded African Americans who listened to *Amos 'n' Andy*. Tuning in to the program, he wrote, was the same as laughing at and insulting oneself. Vann's editorial position did not take Schuyler's critical tone, but even at the start of the *Courier*'s petition drive, the paper editorialized that in taking a stand against *Amos 'n' Andy* blacks declared their respect for themselves—and demanded it from others.[44] And the role that Vann in this case charted for his paper—as a less far-reaching alternative to the flawed mass communication radio offered—had precedent too. This was a familiar role for an African American newspaper, but it took on new significance in an era in which a dominant national mass media did exist.

Social critics such as Vann, Schuyler, Calvin, and Walls do not stand for all African Americans, of course. Several prominent African American newspapers, including the *Chicago Defender* and New York's *Amsterdam News*, opposed the *Courier*'s campaign against *Amos 'n' Andy*. But the voices in the *Courier* represented a perspective on radio at least partially shared by others in the African American intellectual community. The *Courier*'s commentators were not alone in recognizing the power of radio to propagate ideas and images—or in voicing their frustration with the racial images radio disseminated. Although not all, some of the very few radio-related articles to appear in the NAACP's journal, the *Crisis*, echoed pieces of the *Courier*'s attacks on the medium. English professor Ivan Earle Taylor despised radio for presenting the African American only in grotesque stereotypes. "The radio makes him a rogue, a fool or a drunken, care-free bawd," he wrote. And almost eight years after the *Courier*'s protest of *Amos 'n' Andy*, dramatist John Silvera still angrily condemned radio and that program's influence in particular: "Amos and Andy . . . should be nominated for the highest award of the American Society to Keep Negroes Down. In their brief years of existence the characters of Amos and Andy and the fictional Negro they represent have done more harm than ten years of constructive education by the N.A.A.C.P. could remedy. As an agency of propaganda, radio is unsurpassed."[45]

Reflecting on his own experiences trying to write for radio in the late 1930s and during World War II, Langston Hughes concluded that the medium failed America both by perpetuating racist images of blacks and by denying alternative voices a platform to speak. Those in control of broadcasting allowed only familiar stereotypes of African Americans onto the national networks, Hughes observed. And since the narrow oversight of broadcasting's networks meant that blacks had virtually no uncensored access to the airwaves, he saw no way to challenge those stereotypes—and no semblance of free speech on the air. In a vehement letter to radio playwright Erik Barnouw, Hughes charged: "Considering the seriousness of the race problem in our country, I do not feel that radio is serving the public interest in that regard very well. And it continues to keep alive the stereotype of the dialect-speaking amiably-moronic Negro servant as the chief representative of our racial group on the air."

Even supposedly liberal programs failed to overcome the constricted stereotype, Hughes lamented: either a program presented blacks as childlike or it failed to reach the broadcast booth altogether. Hughes had experienced both silencings. And the problem of who was allowed to speak concerned him as much as what it was that Americans heard. More than once Hughes developed radio plays that never aired, even on the comparatively progressive network, CBS. Rejecting Hughes's play, "De Organizer," in 1940, CBS executive Davidson Taylor explained that the network found the play admirable, but too "controversial" to run. As long as the executives running radio sought to avoid controversy and racial questions remained controversial, the medium offered little for African Americans—and for the American public, if understood not as monolithic, but as a diverse whole—Hughes believed. Radio appeared a paradoxical means of communication, one that—in terms of speaking—shrank rather than expanded the number of communicators. During World War II Hughes thundered, "Personally, I DO NOT LIKE RADIO, and feel that it is almost as far from being a free medium of expression for Negro writers as Hitler's airwaves are for the Jews."[46]

Hughes did not write in systemic terms as he critiqued radio. Unlike Rorty, Brindze, and others, Hughes did not dwell upon who controlled the medium and how it was structured. And unlike Rorty and most of those who shared his mass-production critique, neither Hughes nor Vann and his columnists focused their attention on corporate dominance of the air. Radio concerned Hughes and some of the *Courier*'s writers because the medium fostered racial stereotypes and at the same time denied blacks a chance to combat those images. These commentators did not see radio as inherently bad. They, in fact, wanted access to the influential airwaves. Hughes, Vann, Schuyler, and Calvin wanted a greater voice in a national discussion of race. In this respect, as Hughes and the *Courier* called radio to task for spreading racist propaganda, they explained clearly the risks inherent in a mass-produced culture that enabled a select group to make decisions about who could speak and what the nation heard. For those not part of that group, these critics suggested, the rise of mass communication could well threaten to take public speech away from the people.

Related Solutions

Those who critiqued radio from either the point of view of the mass consumption or mass production of culture had no doubt that broadcasting in the United States needed reforming. The prescriptions Orton, Rorty, and like-minded thinkers offered for radio demonstrated both the differences between the two critiques and, fundamentally, the common ground they shared. Ultimately, both camps of thinkers sought ways to break apart the mass public they found so ominous. Both favored experimenting with government action to split mass culture into smaller pieces. But in their divergent proposals for that action, these commentators revealed the different implications of their two diagnoses of mass culture's problems. Some— generally those concerned with mass culture's tendency to dull individuality and distinction—favored state-run broadcasting that resembled Britain's BBC. Others—typically those who saw mass culture primarily as a threat to free speech—proposed using government activism to create alternative stations and broaden access to the air.

America's broadcasting system was so deeply flawed, William Orton and James Rorty agreed, because while staggering material changes had swept across the United States in the days since Edward Bellamy had looked forward and imagined the twentieth century, Americans still retained their traditional values and understandings of the world. Those conceptions and organizing principles and were now severely outdated, critics in the 1930s charged. "Almost every material aspect and activity of life is radically different from what it was fifty, even twenty years ago," Orton wrote. "But our outfit of concepts and techniques for the ordering of our life as a society is still heavy with the dust of the eighteenth century." Radio was obviously crucial among the technological disruptions that recent years had sprung upon the United States. To Rorty, it was a disruption that metastasized beyond society's control: "Radio broadcasting is certainly one of the most disruptive of all our new technological means. It cuts right through the whole context of our obsolete or obsolescent political, legal, educational and cultural concepts, institutions and habit-patterns. . . . Broadly stated, the issue is whether or not a political democracy, holding the bag of an obsolete, unplanned, traditionally exploitive capitalist economy, can pull radio out of that bag and make it approximately functional in the interests of human progress and civilization."[47]

Lack of planning had helped create the problem, both sets of critics believed. The solution, then, was to plan for the government to take a more active role in structuring radio in America. In theory, Orton said, demonstrating his evolving liberalism, he favored letting organizations develop naturally, without the social planning many modern thinkers favored. But, he admitted, that sort of lack of foresight had devastated mass communication in the United States. Radio, he concluded, needed state guidance. Broadcasting, Orton declared, was not an ordinary business, but a public trust. As a source of news and culture, radio was too important to be left to private hands guided only by profit. Rorty and other left-leaning thinkers reached a similar conclusion. With unfettered freedom of the air technically impossible due to limited frequencies, Rorty believed radio was inherently an arena of conflict. "The task of statesmanship," he wrote, "is to make that conflict creative, rather than destructive." Only through federal oversight and planning, he explained, could America profitably balance the forces competing to dominate the air.[48]

The question became, then, what that government involvement should look like. The answer depended, in part, on what aspect of mass culture most concerned a critic, on whether one worried most about what listeners heard or about who was doing the speaking. For technical reasons, Orton believed radio would be most effective under monopoly control. But clearly he opposed a private monopoly of the air. State control of the airwaves, though, offered a solution: government control removed the profit motive that Orton argued led broadcasters to push individual listeners into an indistinguishable mass audience. He and many other critics pointed to England as a model for American broadcasting. The British Broadcasting Company, run by a state agency, controlled all English stations. As a result, many public intellectuals wrote, the BBC produced quality, serious programming without ads. To many thinkers, both liberal and left-leaning, this seemed the best setup. Louis Filler agreed: he saw monopoly control of radio as unavoidable—so America needed to make that monopoly a good one. "Centralization of radio control is inevitable," he wrote. "The only question is: what form will it take? Will it emerge as a vast private monopoly, ready and willing for fascism; or as a public utility, subject to public control?"[49]

To others, the challenge lay not in managing a broadcasting monopoly to serve a diverse public, but in preventing such a monopoly in the first place. To Rorty, Brindze, and a smattering of other critics, what mattered most was maximizing freedom of expression by maximizing who had access to the airwaves. Consequently, Rorty and Brindze favored government intervention, not to run broadcasting, but to prevent commercial monopolies of the air and to increase the diversity of stations and voices. Rorty sought an "orderly conflict" on the air, not tidy control in the hands of any body. The solution, as Brindze saw it, lay in creating a government network to compete with the commercial ones. The government stations would offer minority views ignored by the existing networks. Additionally, she hoped, the networks would improve their programming when they failed to measure up to their new competition; a government station could serve as a yardstick in the spirit of the New Deal's TVA, she claimed. The idea of a public broadcasting system running alongside commercial stations won some favor. Various critics suggested that government intervention to create a network of serious stations could improve news broadcasting and spur existing networks to elevate their programming in general.[50]

Clearly, sharp differences existed between these two reform plans. Brindze worried that a state-controlled broadcasting system offered no improvement, simply substituting one monopoly on speech for another. On the other hand, the *New Republic*'s Bruce Bliven, in calling for a BBC-style system, suggested that quality of programming, not censorship, was the biggest problem facing radio.[51] In general, those who favored government-controlled radio were guided by a fear of a system that constructed a mass audience at the expense of distinct listeners. To those critics who called for greater competition on the air, the more crucial issues at stake were those of speech, democratic participation, and access to power.

But the differences between these proposals—and indeed between the intellectual camps staked out by critics of mass-consumed and mass-produced culture—were not, in fact, quite so stark. With a few exceptions, these thinkers rarely advocated for one solution entirely at the expense of the other. Moreover, in both cases, commentators advocated for government planning to direct what was often understood as private enterprise. That in itself was a significant step. As the New Deal in Washington, D.C., similarly instituted government action in other aspects of life in the United States, Franklin Roosevelt's administration contained the same variety of ideas under its umbrella.

More significantly—whether they favored a state-controlled system or a competitive TVA-style model, whether they approached mass culture from the point of view of consumption or production—all of these critics had the same basic goal in mind: to attempt to use public planning to break mass culture into smaller, more individualized pieces. Both sets of thinkers worried about the creation of a mass public. To Orton, that meant a homogeneous audience that blurred individuals' distinctions. To Rorty, that meant something more abstract, an arena of meaningful civic discussion that had become too large and centralized to allow for ordinary citizens to offer their own views. But both saw radio creating a mass public, and both legitimately feared its implications for the United States. The difference between the two reform plans might be linked to these thinkers' different notions of democracy, whether democracy most properly offered a liberal defense of individualism or served as a means of fostering civic involvement. Here again, a basic similarity connected these views, though. Both sets of thinkers worried about the fate of democracy in a mass culture.

Not all public intellectuals did.

Defenders of the Faith

Even in intellectual circles, radio had its ardent supporters. The new medium appealed to a small swath of the commentators seeking an educated audience because some of them appreciated the prospect of a mass culture. To these thinkers, commercial mass culture inherently empowered, not oppressed, the populace because the effort to appeal to a mass audience made popular taste the supreme cultural arbiter. Implicitly, these thinkers engaged radio's critics in a debate about where power lay in a mass culture.

Almost as fundamentally, radio's defenders and critics also clashed over the very concept of a mass public. Radio's intellectual supporters would have agreed with the medium's opponents that twentieth-century America created the public as a mass entity. But broadcasting's backers fully accepted this modern construct. To this group, a mass public did not threaten American democracy because they understood the ideal of democracy in terms of masses and collections of people. Freedom to choose among various options coupled with majority rule, they implicitly suggested, characterized their ideal of democracy—not individual distinction or personal civic empowerment. A system in which authorities had an incentive to seek the approval of the largest chunk of people possible ensured that the will of the masses would prevail. In other words, precisely because radio was a form of mass culture, then, intellectual supporters of radio saw broadcasting protecting the consumerist virtues of a mass democracy that they so valued. Rather

than seeking to break down the mass qualities in their new culture and rebel-
ling against a world in which individuals counted principally in aggregate,
they embraced their mass culture and its implications.

To radio's intellectual defenders, then, mass culture was neither degrading nor dangerous. It only appeared so to those who mistrusted the mass
of Americans, they implied. Radio in the United States, these supporters
claimed, staunchly championed their democratic ideals. The mass medium
democratized culture, spreading it to the common people, and in programming, the desires of the majority, not those of a distinct few, ruled. The
commentators who articulated this stand tended to be more conservative
politically than the medium's critics: they almost never wrote for the *New
Republic* or the *Nation*; and some vocally opposed the New Deal. More significantly, they tended to be involved in broadcasting themselves.[52] But their
ideas still reached a large and well-educated audience. And to that audience,
they argued that radio had a crucial role to play in modern America. Radio's
defenders argued that their chosen medium offered a cultural boon for society as a whole and served as a bulwark defending the will of the people.
Commercial broadcasters had a financial incentive to program the airwaves
according to the majority of the people's tastes, supporters claimed. Moreover, keeping radio out of the government's hands reduced the risk of state
tyranny. This hardly amounted to a highly sophisticated intellectual defense
of the status quo. But on the eve of World War II, as state fascism loomed
ever more threatening in Europe, the pro-commercial-radio position became
increasingly compelling. As radical and liberal intellectuals turned their ire
against the Nazi state, some critiques of private broadcasting lost a bit of
their edge.

Those who defended radio's virtues in intellectual arenas frequently
trumpeted the musical potential of the medium. These commentators often
shared Orton's sense of cultural hierarchy. But unlike the critics of culture for
mass consumption, radio's defenders believed radio did not undermine that
order; rather, broadcasting benefited society by helping to bring high culture
to the masses. To a handful of conductors, composers, and critics writing
for an educated audience, radio could expose new listeners to rarified and
elevating classical music. The medium still had technical flaws, composer
Leopold Stokowski wrote, but it did offer "a means whereby Everyman can
hear music with overwhelming beauty and eloquence in any part of our vast
country, no matter how remote."[53]

This was not, however, an argument to let elite tastes govern programming. It was equally an assertion of the importance of typical listeners and
their values. Composer and music critic Deems Taylor went so far as to suggest that the prevalence of unsophisticated music on the air actually helped

the cause of disseminating classical and symphonic numbers: if people tuned in for popular music, he reasoned, they might accidentally hear a few serious pieces while listening. These thinkers had no doubt that radio could educate and spread uplifting culture to the masses. Quite simply, the prominent popular culture critic Gilbert Seldes claimed, radio had democratized serious music: it was no longer a luxury item accessible for only the wealthy few.[54]

For radio's supporters, the popular quality in American broadcasting was crucial. The commercial nature of radio made the network system responsive to the will of the people, these commentators maintained. American mass culture was, in this analysis, subject to the populace. The people as a group ruled the American airwaves, pro-radio thinkers believed; to win customers, advertisers had to please not a few individuals, but the masses. "So long as radio is used to sell things to us," Seldes wrote, "we have that degree of control." American radio, these critics argued, had to meet the public's desires because listeners could effortlessly tune out otherwise. Consequently, supporters cheered, broadcasters played to the people's tastes like no entertainment before. The radio industry eagerly sought to determine what listeners wanted and to give that to them. Taylor explained that "radio as it exists in this country is one of the most completely democratic institutions in the world. The broadcaster, far from ignoring vox populi, is desperately, pathetically anxious to catch the sound of vox populi. If he fails to hear it as distinctly as we think he should, it is probably because many of us do not take the trouble to speak loudly enough. In the democracy of radio, as in the democracy of politics, the so-called 'better element' stands on the sidelines and complains, while the shirt-sleeve populace goes out and votes."

Those who complained that radio did not meet high cultural standards got what they deserved, suggested Taylor and others. Broadcasters designed programs to meet listeners' preferences; and elite listeners seldom wrote to stations to express their desires, while popular listeners frequently voiced their views.[55]

In other words, radio reflected the tastes of the masses who registered their opinions. To broadcasting's intellectual supporters this amounted to democracy at its finest—and hence radio working as it should. "While the arrangement may not result in great literature or the most cultured programs it is democratic and, as one looks round the world to-day, anything democratic is worth fighting to protect," wrote playwright Merrill Denison as fascism spread through Europe. To Seldes, any other system smacked of authoritarian rule. He recognized critics' laments about the low quality of radio programming, but believed that those thinkers had missed the crucial point at hand—and missed it by a dangerously wide margin. "Approaching the major problem," Seldes wrote, "would it be better for a democratic people to have good programs whether they asked for them or not, than to have programs of mixed good and evil elements in which they can eliminate the evil?" Seldes found but one possible answer: the mass of the people was and should be the dominant group in American society; mass tastes and opinions should reign. Anyone who took a different stand, he hinted, turned away from democracy as he understood it. "That a mechanism capable of reaching all the people should attempt to appeal to all the people, and doing so, should fall below a cultivated taste and a rigorous intellectualism is surprising only to the most abandoned of idealists," he wrote, chiding radio's opponents. "That it should not be controlled in the interests of aesthetes and against the interests of the dominant class, is irritating to fanatics."[56]

This argument suggested that mass culture was desirable, at least in part, precisely because it was culture designed for the masses to consume. These commentators believed in embracing America's new culture, not resisting it. As long as the people—or, more accurately, the bloc of people that made up the majority—counted in the broadcasting system, the idea of trying to break a mass culture into smaller, more individualized pieces was, at best, contrarian. To radio's intellectual supporters, then, the mass production of culture too was a virtue. Only by mass-producing culture could America create a culture for a large cluster of the population. But these thinkers also noted a risk here. In the wrong hands, a centrally controlled culture might not serve the interests of those masses.

Radio's defenders took that danger seriously, but unlike the medium's critics, they saw no sense in breaking the mass medium or its audience into smaller pieces to alleviate the threat. They wanted a system that insured the rule of mass interests; they applauded therefore commercial radio for giving those in command of the airwaves a financial incentive to meet the majority's desires. The very survival of America's democratic political and social system depended on the commercial broadcasting system, radio's supporters alleged. No other system—certainly no system of governmental broadcasting—could be trusted. In part, these commentators looked at Europe and saw countries where the rise of fascism had accompanied state control of broadcasting. Radio, these thinkers believed, could bring public issues to the masses and involve listeners in the democratic process like never before. But, they warned, that could take place only if broadcasting were controlled by those with a vested interest in doing so. Radio had such incredible potential as a propaganda machine that, if government officials controlled radio, those in power would almost invariably use it to secure their own ends, effectively shouting down anyone who disagreed. "It is nonsense to say that radio is necessarily an agency for civic good," journalist William Hard wrote in a comparison of radio in the United States and Europe. "Radio, monopolistically controlled for the purposes of persons in power, can be made the most effective agency ever devised for the enslavement of the mass mentality of a nation."[57] Only by keeping radio in private hands, these analysts believed, could America prevent the medium from supporting dictatorial forces; political bodies, they feared, could not govern the air impartially.

Obviously, radio's defenders failed to see that business leaders also had their own interests and used the medium for their own propaganda. But in the face of growing Nazi power in Germany, they had plenty of compelling reasons to focus their attention on the threat of a totalitarian government. As events in Europe unfolded over the course of the 1930s, Merrill Denison came to regard the commercial broadcasting system more and more highly. In 1934, although he did not support government control of the air, Denison blamed the existing system of sponsorship for poor programming. By the end of the decade, though, he faulted American radio for having too much government involvement and claimed that only private control of the medium could support freedom and democracy.[58]

Radio's more conservative supporters were not the only public intellectuals to see value in American broadcasting as the Depression crisis faded into the threat of world war. By the late 1930s radio's radical critics periodically and grudgingly acknowledged that American radio might empower people in some small ways. As radio gradually proved its worth as a news organ over the decade, and particularly during crises in Europe, some critics of the mass-produced medium decided broadcasting did help keep listeners informed and involved in events—and therefore better able to participate in the political process. T. R. Carskadon of the *New Republic* declared American radio "commerico-moronic" in 1937. But, he admitted, broadcasting did have its very rare "moments of pure fire": no other medium could bring live, newsworthy events or speeches to the masses as effectively. In the aftermath of the 1938 Munich crisis, the *Nation* suggested that radio's news coverage of the brush with war was an achievement worth celebrating: for the first time in history ordinary people could attain almost instantaneous updates on the progress toward war or peace. Even James Rorty agreed that American radio had shone during the crisis, helping to involve listeners in the events that governed their lives. "A new dimension has been added to politics and diplomacy," he wrote. "For the first time history has been made in the hearing of the pawns." Rorty still feared commercial broadcasting, but state censorship of European radio in the late 1930s and American radio's coverage of the international turmoil convinced him that the commercial system conveyed news to the people more honestly than any existing alternatives. Radio in the United States was not free, he wrote; it was, however, closer to free than any other system in the world.[59]

Critics such as Rorty did not entirely forsake their attacks on the mass medium as international events raised fears of an undemocratic state controlling society. But the political climate of World War II understandably made it harder to point to capitalism as the leading source of totalitarianism; and the cold-war conservatism that followed further discouraged such stands. All of this suggests that radio's intellectual defenders' ideas had staying power beyond the originality of their insights. On the surface, these public thinkers did little more than accept the idea that private enterprise promotes freedom, and apply that notion to the new mass communication. They essentially argued for uniting consumer choice and political freedom. But the endurance of radio's defenders' ideas went deeper as well. In embracing the mass medium, these commentators embraced or at least accepted the mass world they were coming to inhabit. American radio, these commentators believed, could spread culture to the masses and enable those masses of Americans to shape culture by turning their dials. This was, radio's defenders argued, more than laudable: mass culture was fundamentally democratic because it implicitly endorsed a democracy defined by majority rule, a democracy in which not individuals but large groups counted, a democracy for a mass society.

We might well balk at this idea: should modern democracy exist only in aggregate? In fact, for all their seemingly populist leanings, those public intellectuals who praised radio in the 1930s missed a crucial point—about living in their modern world and about what the very populace whose tastes they so prized actually desired. Radio's listeners did not want to count in their mass culture only as members of large abstract blocs. Listeners still wanted to matter on an individual basis.

William Orton's consumerist evaluation of mass culture and James Rorty's mass-production critique both highlighted the challenges the new mass culture posed to that goal, to maintaining individual autonomy in a society that felt increasingly large, centralized, and impersonal. But in their assessments of radio, both also largely left ordinary individuals out of their analyses. Orton and like-minded critics essentially suggested that programming determined the medium's meanings. And Rorty, Brindze, and their fellow commentators argued that the industry's structure defined how the mass culture impacted America—a point several African American thinkers vaguely affirmed.

To those who focused on mass culture as something designed for mass consumption, radio seemed to blur distinct individuals into an abstract and impersonal mass. Modern mass culture threatened to transform many small publics into a singular mass public. Mass culture might push an overwhelming and impersonal world into individual lives. To those who emphasized the production side of mass culture, radio appeared to centralize cultural control in the United States. They questioned how ordinary people could meaningfully enter into the broader world and speak publicly in modern America. Together, these two notions—the mass production and consumption of culture—neatly described the culture that radio and the twentieth century wrought. And together, these two sets of critics rightly identified crucial difficulties those changes brought. In the world radio helped to make, Americans would have to face challenges to their efficacy in their own lives and their ability to be heard in the public sphere.

But, again, like radio's supporters among public intellectuals, its critics overlooked what radio actually meant to the American people. Listeners did not simply accept becoming indistinct pieces of a mass culture. But neither did they experience radio as completely undermining their ability to function as individuals or choking out their voices. At the center of the debate, radio's intellectual detractors and defenders disagreed on how people participated in mass culture: whether mass culture is largely irresistible or whether it essentially conforms to the will of a mass audience. The reality—which public intellectuals of the 1930s missed—lies somewhere in between. Radio's critics were right that broadcasting helped significantly limit listeners' control, but those listeners also used the medium to open up new possibilities. In fact, as we will see in the chapters that follow, many Americans came to see radio as a way to help them address the very problems mass culture raised.

Orton, Rorty, and their fellow critics, it turned out, asked many of the right and fundamental questions about their age—even as they proved to be wrong about some of the answers.

2

RADIO'S LISTENERS

Personalizing Mass Culture

As much as fifteen-year-old Janet Bonthuis wanted to give her father a Christmas present in 1939, she knew her Depression-wracked family could not afford to exchange gifts. Hard times demanded sacrifices. Janet, however, had not lost faith in Christmas magic: in preparation for the holiday, she wrote a letter asking for a tin of tobacco to give her father. Several weeks later, she found her dreams answered. And on Christmas day she surprised her father with the present. But this is no simple yes-Janet-there-is-a-Santa-Claus tale. Janet had not addressed her letter to the North Pole. She sent it to a radio program.

In late November 1939, when Janet Bonthuis began preparing her Christmas celebration, she wrote to a national radio quiz program, *Vox Pop*, for help. Illness confined Janet to a tuberculosis sanatorium in Muskegon, Michigan, twenty miles from her family. For the holiday, Janet's father made plans to visit his daughter. Suffering from the Depression, the elder Bonthuis could not afford gas for the trip. To save money for the fuel, he gave up smoking. That choice inspired Janet; she desperately wanted a gift to give her father in return for his visit. On her own, though, she could not afford a present. And she did not find help in her personal circle of relatives, friends, and acquaintances. So Janet looked to one of her favorite radio programs. No doubt she specifically asked *Vox Pop* for help in large part because Kentucky Club, her father's favorite tobacco, sponsored the program. But it was because she saw nowhere else to turn that she sought out a voice on the air at all. "I'm going to ask you a favor, making all of my Christmas," Janet wrote in her plea: "Isn't there any way I can work or do something to get a can of K. C. tobacco for Daddy's Christmas present? Each week I hear you giving tobacco to any body who answers a question, isn't there some way I can earn one? . . . The tobacco would be my only present to give away. I know how happy Daddy would be. So if you think of a way, I would be so glad, but if I can't, only me will be sorry, because my Dad doesn't know of this."

Come Christmas day, of course, Janet had no need to be sorry. *Vox Pop*'s lead announcer, Parks Johnson, sent Janet a large tin of tobacco, and helped her make the most of the holiday. The program had come through for her personally, and she voiced a deep, personal gratitude. "Gee you surely are good," she wrote Johnson thankfully. "I hope you have just the most swell Christmas. Because you've given me the chance to have a nice one, and I want you to know, I send a great big thanks. and if ever I can, I'd like to do something for you."[1]

Radio clearly occupied a special place in Janet Bonthuis's life in the 1930s. When she could not manage on her own, when her local and personal community proved inadequate, she reached for an ethereal connection. Through radio she found ways to personalize a vast, anonymous world. She heard a human voice attached to a large commercial company—and believed it might hear her. The broadcaster who spoke to millions felt like an intimate friend, someone she might trust—perhaps like someone who might help her find a measure of control in her own life.

Janet was not alone. In the 1930s, millions of Americans felt themselves struggling in a world beyond their command. They felt an abstract broader world intruding upon their daily lives, threatening to swamp their traditional personal moorings. The Great Depression helped make it plain that the reach of modern society exceeded the sphere ordinary Americans could grasp through face-to-face experiences. Authority was increasingly distant, and individuals experienced it in increasingly standardized ways. But like Janet, millions of Americans found an answer in radio. They would use the leading expression of their new mass culture—broadcasting—to help them find resources to navigate within that culture. Using radio, Janet and others discovered, they could remake their new public relationships to resemble their private ones and, in so doing, they could personalize and function in their impersonal mass culture.

Nothing, of course, about the technology of radio or the administration or structure of the medium in the 1930s dictated that broadcasting would help individuals uncover a sense of self control as they increasingly interacted with the widening world. Looking at the America he saw emerging in the 1930s, economist William Orton wondered if it would still be possible to function as an individual as modern mass culture took hold more and more firmly. He and other public intellectuals were right that radio was a crucial source of that culture. Since late in the 1800s, the United States had been slowly growing more interconnected, centralized, and standardized. By the beginning of the Depression, radio brought those same changes to millions of Americans by transforming the very air, making culture a mass experience.

What those critics ignored, however, was what audiences who listened to radio's popular programming did with what they heard on the air. Radio mattered enormously to many of those who heard it in the 1930s; their experiences with and understandings of the medium deserve attention. Although broadcasting represented a powerful form of centrally created and uniform culture, the programs and structure of radio did not just impose a meaning on listeners. Popular listeners did not dictate the shape of radio or its programs, but within the bounds of the centrally controlled form, those listeners discovered some room to use radio in ways that helped them to count in modern society. The advertisers who controlled radio programming, for instance, often treated the audience as a mass of impressionable and uniform consumers. The critic James Rorty was right that radio forcefully distributed consumerist values. And it was true that listeners often accepted and internalized those values. But at the same time, listeners also occasionally used those consumer identities to help them gain voices that might be heard faintly by the distant and abstract forces that controlled mass culture. Listeners were not simply manipulated by their mass culture, but found limited possibilities to negotiate within it.

Listeners like Janet used that space and the mass medium they so valued in order to help them face the challenges the expanding mass world posed to their autonomy in the circles of their own lives. As listeners found their local resources insufficient, they used radio to help them reimagine the new mass world in personal, familiar terms; and they drew upon that new sense of community for valuable sources of support. Listeners crafted new kinds of relationships with the people and fictional characters they heard. They used radio to reinvent abstract public relationships as personal relationships. They successfully imagined that the new mass communication of radio resembled older, more obviously intimate, forms. And they befriended the voices they heard through that medium.

In forging connections with radio voices, many listeners blurred the line between the public and the private in their lives, enabling them to find in the public mass medium of radio the same types of support they once might have found in face-to-face communities. In their relationships with figures on the air, listeners found information, personal advice, connections to far-off authorities, and expert guidance in a world that defied the limits of traditional resources. These modern ties lacked the concreteness of geographically based ones to be sure; but in the vast, interconnected society of the twentieth century, listeners like Janet Bonthuis came to depend on such socially constructed communities.[2]

The Mass Audience Listens

Had Janet Bonthuis lived through her Christmas in the sanatorium even a decade earlier, she almost certainly would not have written to a radio program when she needed a gift. Radio first took on a real importance in the majority of Americans' lives in the 1930s. Because so many Americans came to care so deeply about radio at that point, it was during the 1930s that radio first attracted a mass audience. The ways audiences listened—their access, what they heard—played a crucial role in making broadcasting a mass experience in the Depression decade. New listeners invested themselves in broadcasting in droves in the 1930s. National audiences of a spectacular scale coupled with the growing reach of networks which standardized what the audience heard combined to make radio America's dominant mass medium. That does not mean, of course, that all Americans experienced exactly the same culture in exactly the same ways. Radio's audience included a huge portion of the country—crossing many, though not all, bounds of geography, ethnicity, class, and gender. Not everyone listened the same way or heard the same programs, but a mass audience need not be entirely inclusive. Moreover, as particular popular genres developed in the 1930s, huge numbers of Americans did tune in. And many responded in common ways. Listeners interacted with the programs they heard. That interaction represented radio's importance to millions of them.

On the simplest level, the creation of radio as a mass medium grew from its accessibility. For the first time in the 1930s, the majority of Americans had access to broadcasting. The price of radio receivers plummeted during the decade—even as the quality of the still new technology jumped. In 1931 RCA's least expensive receiver cost $37.50; a decade later one could buy an RCA radio for $9.95. Ten years after the first scheduled radio broadcast in 1920, radios were known throughout the country, but had not yet found their way into the majority of homes: 40 percent of households in the United States had radios in 1930. In the decade that followed—despite the economic crisis—radio became commonplace: 86 percent of homes had sets by 1940.[3]

Radio would attract a mass audience because those millions cared deeply and made broadcasting an integral part of their daily lives. By the end of the 1930s, Americans ranked listening to radio as their favorite recreation in a

Fortune magazine survey. And when asked which they would be more willing to do without, going to movies or listening to radio, those polled favored
ditching the movies to losing radio by almost six to one. Small wonder then,
that some social workers reported that impoverished families would give up
furniture or bedding before selling a radio. "If you took the radio away from
[my ten-year-old son] it would be like taking away his meal," said one unemployed man. "If we lost the radio, I'd be willing to dig my own grave," said
another.[4] What they heard on the air mattered to listeners. That meant radio
kept its mammoth audiences' attention. As baseball's Detroit Tigers rushed
toward a spot in the 1934 World Series, fans in Michigan and nearby states
called for NBC to use Detroit announcer Ty Tyson to broadcast the series.
In just three weeks, organizers of a pro-Tyson petition drive garnered six
hundred thousand signatures. Baseball's commissioner vetoed the possibility
of Tyson announcing the games, but NBC executives were so impressed by
their listeners' fervor that they set up two broadcasting teams: the baseball-
approved broadcasters announced the game over NBC's national network,
while Tyson could be heard on the Detroit NBC station.[5]

Despite NBC's sleight of microphone, most fans listening to the World
Series heard the same actions, described by the same announcers, at the same
time all over the United States. The development of national network broadcasting in the late 1920s meant that listeners across the country could hear
the same programming, the same events, the same ideas. The mass audience
received a standardized, mass-produced product. In the wake of the federal government's 1927 reorganization of the airwaves, centralized national
programming came to dominate. In 1937 New York-based radio networks
controlled 93 percent of America's broadcasting power. And 88 percent of
listeners said they preferred network programs to local ones. Although local
stations existed in many cities, the vast majority of American listened mostly
to offerings created by either NBC or CBS and distributed nationwide. Listeners found it easy to be a part of a common national culture.[6]

That national culture, it should be noted, did not include the whole of
the nation. Even in the age of radio, not everyone was included equally as a
member of the common audience. A mass culture is rarely completely inclusive; access to the expressions and values it transmits in fact represents a
limited form of power. While radio crossed many divides in building a mass
audience, it did not do so absolutely. To some degree, for instance, radio's
uniform audience was constructed around middle and working-class listeners. How people listened varied somewhat—though far from absolutely—by
class. Residents of households making between two and three thousand dollars a year were one-third more likely to listen to the radio in the evening
and twice as likely to listen in the day than residents of households with an

income greater than five thousand dollars a year. The same study revealed that, to some extent, different economic classes listened to different programs. Wealthier listeners preferred programs of opera and symphony, while those less well off tended to tune in to serial dramas, comedy, and variety programs more frequently. As advertisers sought to maximize listenership, programs that network executives saw as pleasing to working- and middle-class tastes dominated the airwaves and those groups came to define the popular audience. Radio then created a mass culture rooted in certain class-based appeals; not everyone felt included equally in radio's audience.[7]

The gaps in the mass audience were starker, though, in geographic and racial terms. Regional distinctions persisted even as radio broadcast culture nationally. The majority of Americans owned radios by the end of the Depression, but even then those radios were not evenly distributed through the country. Rural southerners acquired radios far more slowly than urban Americans or even farmers in other parts of the country. In 1935, when an estimated 81 percent of the homes in the Northeast and 77 percent of the homes on the Pacific coast had radios, only 62 percent of homes in the Midwest and just 48 percent of homes in the Southeast had receivers. In Mississippi and Arkansas radios were scarcer still: only about a quarter of all homes possessed sets. Few of the working-class southerners interviewed by the Federal Writers' Project in the late 1930s for *These Are Our Lives* included listening to radio among their amusements. Like others interviewed, one older woman named Lizzie often found herself sitting in bored silence: her eyesight made reading hard, she said, and her family could not afford a radio. This meant the importance of radio was not entirely homogeneous. Farmers in Oregon, for instance, vehemently supported local agriculturally oriented broadcasts. Comparatively, when North Carolina's agricultural department gave up its radio program in the mid-1930s, the department did not receive a single objection. Not surprisingly given the demographics, radio access also varied by race. African American ownership lagged far behind the national average: in 1930 only 7 percent of African American families had radios; whites were six-and-a-half times as likely to be so equipped. Radio became much more widespread over the 1930s, of course, but even late in the decade receivers remained disproportionately concentrated in the hands of urban and nonsouthern whites. It was at least 1938 before a majority of black homes had radios, for instance.[8] Those left out of the cultural producers' target audiences remind us that some capital comes with being included. The composition of the mass audience was complex, not absolute.

In its complexity, though, a mass culture existed; and, whether or not someone had a radio, the medium often reshaped how he or she inhabited the modern world. Not owning a radio, for instance, was not exactly the same as not hearing a radio. Want-to-be listeners proved resourceful. Those without sets occasionally listened with neighbors or in public places such as stores. Like other radioless farmers, one Georgian reported traveling for miles to catch a particular broadcast. "I walked four miles and forded two streams just to hear your seven o'clock program," he wrote to an Atlanta station. And, in his nostalgic memoir, Clifton Taulbert remembered crowds of African American listeners gathering around one of the few radios in his small Mississippi town to hear Joe Louis fight. Farm workers made the trip into town to crowd around Taulbert's family's radio, overflowing the front porch and sitting on the ground. Some other southerners recalled turning to grapevine radio, in which families who could not afford a set or lacked electricity might run a long wire to a neighbor's set and connect a speaker to it, listening in to what the neighbor played.[9] Of course, in all those cases listeners had little say over what they heard or when they heard it, but they did experience the common medium.

The importance listeners clearly attached to radio and the meanings they found in it stemmed, in part, from the ways they listened and what they listened to. In general, modes of listening in the 1930s brought the outside world into familiar surroundings. Whether listeners tuned in on large sets that were almost as much furniture as receiver or on smaller models, their sets often occupied a favored spot in the house or apartment. "Due to its place in the home," said one ad, "a new radio becomes one of the most important pieces of home furnishings."[10] Families often listened together. It would ebb as radio spread over the decade, but the idea of listening as a group activity was established by the start of the Depression. In 1930 surveys found four or more listeners clustered around a typical set at any time. Even in the late 1930s, studies generally found between two to three listeners for each set during the evening. Only on the eve of World War II did group listening seriously decline as more and more homes could afford radios, and many families began purchasing multiple sets.[11] Only then did radio really begin to break free from the home—but even as the Depression faded, listening often took place in quasi-private space. The first radio in a car debuted in 1930, but that did not become common until the end of the decade: it took until 1938 for the total number of car radios to reach five million.[12] Moreover, listeners found many different ways to listen in their homes. Listeners readily used radio as either the centerpiece of their entertainment or to supply background sound for other activities. At times listeners paid close attention to the airwaves; other times they might play cards, do chores, or read while the radio played.

Those listening styles often varied by gender. Not surprisingly, since listening typically took place in the home, women listened more often then men in the daytime. And women were more likely to do work while they listened, to use radio as a backdrop to other activities. In the mid-1930s each Sunday afternoon A. D. Medoff of Lambertville, New Jersey, sat down and tuned into *Vox Pop* while his wife cooked dinner. Periodically, he called her in from the kitchen to hear a particularly funny snippet; that distraction cost the Medoffs more than one burnt dinner. Several weeks after losing a roast tongue that way, Medoff wrote, he again listened raptly while his wife worked in the kitchen:

> She was watching the french fried potatoes that were being fried with fat. At my request, she came in to listen with me. A few moments later, the odor of something burning presented itself, and my wife ran into the kitchen to investigate. The fat in the french fryer caught afire. Of course, she yelled for me to come in, but I remained listening to the radio, thinking that some more meat burned, which I was sort of used to by now. When I did not come in, she started to scream. I quickly got up and ran into the kitchen. Upon entering it I noticed flames of fire shooting from the pot. By this time, the cupboard which is adjoining the stove, and the wall were starting to catch fire. I grabbed a towel, caught hold of the handle of the pot, and threw it on the back porch, but not after singeing my right arm.
>
> While on the porch, the pot continued to burn, and the flames started to go higher and higher. Not having any experience with extinguishing fires, I filled a dish with water and threw it on the flame. The moment the water came in contact with the flame, there was heard an explosion, and only by some quick footwork did I manage to stop my clothing from setting afire. Needless to say, we did not have potatoes for dinner that day.

One study found that only 13 percent of women devoted their full attention to programs during the day, while 79 percent did so in the evening. During the day, radio competed with work in the house for many women's attention; at night, women's listening more closely resembled men's.[13] In their gendered listening patterns—though perhaps not in their fire safety skills—the Medoffs were fairly common.

Despite the significant differences in how they listened, men and women were alike in making radio an important part of their lives. Radio was a gendered medium—from access to the airwaves, to the narratives conveyed on the air, to some of the ways audiences listened—but as we will see, men and women shared many of the same basic understandings of what radio meant to them. For both women and men listening was an active process: both groups interacted with what they heard, wrote letters to programs, and, ultimately, crafted meaning out of the airwaves. Moreover, in a broad sense, the audience that radio constructed included both men and women. Men and women alike participated in the same mass culture. Overall, they enjoyed similar types of programs in the 1930s. There were some differences in taste: men liked sports broadcasts much more than women, for instance, and women listened to daytime soap operas more than men. But simply because women dominated the daytime audience, that does not mean they preferred entirely different types of programming. As media historian Michele Hilmes suggests, many popular evening programs used the same format as daytime ones. Significantly, most of the evening hits appealed to both men and women alike.[14]

Musical programming, for instance, attracted a broad audience. Throughout the 1930s music remained the most-played genre on the air. The amount of music broadcast declined through the Depression, but even on the eve of World War II it was a tremendously popular format. Audiences who had recently had to make their own music or find a way to access to one of the period's uncommon and limited phonograph collections suddenly found music available virtually whenever they wanted it. And listeners loved that. One listener appreciated his newfound ability to hear music at the twist of a knob so much that when his radio broke, he parked his car under an open window and played the car radio loudly enough to dance in his house.[15]

No single all-music program, however, drew the audiences of the most-popular hits. The most listened-to individual programs tended to be serials or comedy/variety programs. The first colossal radio hit, in fact, was both a serial and a comedy. When NBC first aired *Amos 'n' Andy* in 1929, America discovered just how significant a role a program could play in listeners' lives. The story of a listener hearing an episode of *Amos 'n' Andy* without missing a line simply by listening through open windows while walking down a street has been repeated so often that it seems apocryphal. Like other popular programs that followed it—though to a greater degree—*Amos 'n' Andy* infiltrated American culture: it spawned a comic strip and candy bar; movie theaters had nightly intermissions to air the program for patrons; at one mill, workers convinced their employers to change shift hours so they could hear the program; phrases from the program such as "check and double check," and "holy mackerel" crept into common speech—Louisiana senator Huey Long borrowed his nickname, the Kingfish, from a prominent character.[16]

The incredible popularity of *Amos 'n' Andy* faded quickly—by 1932 it had relinquished the top spot in listenership rankings and a few years later it fell out of the top ten—and a range of programs came to shape the listening experience. The comedy/variety format, with programs such as those hosted by Eddie Cantor, Jack Benny, Ed Wynn, Edgar Bergen, Fred Allen, and George Burns and Gracie Allen, dominated rankings through the decade. Serial stories, programs like *The Goldbergs* and *The Aldrich Family* that followed their characters from episode to episode, proliferated and drew large audiences. By the end of the decade, NBC and CBS aired sixty-one different daytime soap operas, one of the most common types of serial stories in the late 1930s. Other genres captured listeners' attention: dramas and movie remakes, amateur talent displays and quiz shows, talks by flamboyant speakers and, increasingly as World War II approached, even newscasts. By 1939 roughly 70 percent of radio families considered broadcasting their most important source of news.[17] Listening to this range of programs, an enormous audience came to value radio more and more.

The process of investing themselves in the new medium was not a passive one for many listeners. Integrating radio into their lives meant interacting with the broadcasts they heard. For many listeners, that meant writing letters to programs. Because of that, those letters reveal listeners explaining, for themselves, the meanings they found in radio. True, not all listeners wrote letters to radio programs. But a great many men and women did; sending a letter to a voice on the air was a common and accepted act. Those voices and the producers behind them worked hard to encourage listener mail, to suggest the legitimacy of such correspondence. With methods of measuring listenership imperfect in the 1930s, the radio industry relied heavily on listener mail to prove a program's popularity. Most programs solicited mail through invitations to write, through contests or clubs. In 1939, for instance, 71 percent of NBC's commercially sponsored programs offered at least one premium of some sort. And listeners wrote in in staggering numbers. Correspondence between ordinary individuals and public figures—whether politicians, entertainers, journalists, or otherwise—skyrocketed. To take one example, in the decade from the mid-1930s to the mid-1940s Mary Margaret McBride, host of a daytime magazine program, received approximately 3.5 million letters from listeners.[18] Most listener letters have not survived, but occasionally particular individuals chose to save that correspondence. In those letters, we can catch listeners' voices.

The complex mass audience that listened so eagerly to network offerings began working out the meanings of broadcasting, and their mass culture more generally, in the 1930s. Through their interactions with popular programs, listeners uncovered ways to confront some of the issues they felt most pressingly. Those listeners found in the medium very limited ways to speak in conversations originating beyond their face-to-face experiences and assistance that they came to rely upon in their own lives as well.

Consumer Bargaining

When the popular comedian Jack Benny took to opening his radio program by greeting his audience, "Jello, everybody," he did more than simply turn his sponsor's name into one of radio's most common joke forms, the pun. In the first lines of his program, Benny expressed the vital connection between consumerism and many popular programs: on the air, advertising and enter-tainment were often inseparable. Those critics of radio who suggested that the mass medium gave its commercial controllers a means to impose values were right. But even as listeners generally accepted the consumerist messages they received, they also tried to use radio's consumerism to find small ways to count within their mass culture. When Janet Bonthuis petitioned *Vox Pop* for a tin of its sponsor's tobacco, she made sure to mention how much her father appreciated Kentucky Club, and how disappointed he would be if he had to smoke another brand.[19] In doing so, she implicitly accepted the radio execu-tives' ideas that entertainment existed to convince listeners to buy. But at the same time, Janet was playing up her status as a potential consumer choosing among products in an effort to influence one of the companies that controlled radio. Like many others, she discovered limited space to bargain within the powerful system in order to find at least a whisper of her own voice.

By the early 1930s at the latest, corporations had firmly yoked the air to the production of profits, using their control of the medium to foster con-sumer values. Broadcasting spread the gospel of consumerism, dramatically reinforcing ideals that had been rising in the United States since the late 1800s. You are what you buy, the good life exists in goods, people are pri-marily purchasers, advertisements suggested. Advertisers dominated the air-waves. Early in the decade, a listener could expect to hear a sponsor's name ten to twenty-five times in a typical half-hour commercial program. In fact, those who controlled the air often referred to the programs themselves as commercials. To some station and network executives, radio was little more than a means of personalizing and improving the sales process. Programs existed to meet the sales goals of their sponsors, advertising executive Nate Tufts, who oversaw *Vox Pop* on behalf of the sponsor, reminded announcers Parks Johnson and Wally Butterworth: "Whatever we do in connection with Vox Pop is primarily not for the [advertising] agency, or for Parks and Wally, but for the products." Listener or creative priorities counted only indirectly in that consumer vision. No wonder radio humorist Will Rogers, then work-ing for Gulf Oil, quipped, "I don't know whether I am any good until I see whether they have sold any gas or not."[20]

To a striking degree popular audiences accepted the commercial role of radio and their own identities as consumers in the broadcasting universe.

Those public intellectuals such as James Rorty and Ruth Brindze who lamented the cultural power that big business secured from their hold on radio mapped the landscape accurately. The mass medium forcefully imposed a consumer outlook on listeners; the ideas of America's mass-produced culture proved very hard to reject. Listeners came to equate people and entertainment with the goods they advertised. According to several studies, listeners overwhelmingly knew which companies sponsored popular programs, and sponsorship shaped listeners' buying habits. Often, listeners consciously decided to buy advertised goods to support a well-liked program's sponsor. To many members of the audience, the advertised goods were not simply products, but expressions of a popular radio personality; some listeners saw consumption as a means of being close to and showing appreciation for a radio hero. Fans of the soap opera *The Story of Mary Marlin* such as Chicago's Joy Cole made a point of writing to the program's sponsor, Proctor and Gamble, to thank the company for producing the program: it was proper to repay the sponsor for the entertainment, Cole explained. Other listeners did not distinguish between the product and the program even that much. Many letter writers accepted the industry's nomenclature and referred to radio shows as programs and commercials interchangeably. When one listener wanted to praise Alois Havrilla for his singing, she included a compliment for the sponsor, as though Havrilla and Campbell's Soup shared more than a performance contract: "Your program is the *best* in the air just like your soups are the best in the market," wrote Mrs. Pete Detzal of Erie, Pennsylvania. This consumer connection also worked in reverse. Upset that *Vox Pop* had not returned a photograph she sent to the program, Patricia Pierson of St. Louis expressed her anger by insulting the sponsor's product: Molle shaving cream, she said, was greasy and smelled awful. In either case, the lines between the listener's relationship with the entertainment radio provided and the consumer goods it advertised were fuzzy. To a real degree, listeners united consumer values and listening. From Long Beach, California, George Graham wrote that "Monday nights are reserved, in our family, for your very interesting broadcast of Vox Pop and much irritation, such as that which is felt by the face BEFORE the use of MOLLE [shaving cream], is experienced whenever an interruption is made during the time you are on the air," indicating the centrality of the consumer message to his listening experience.[21]

Accepting the commercial nature of American broadcasting, however, did not make listeners passive lumps of clay, molded entirely by what they heard. Listening to the radio was an act of negotiating between the power of the mass medium and one's own interests. Listeners at times used their status as consumers to assert that their views should count in the vast world.

Occasionally, listeners suggested that program producers should pay particular attention to them because they used a sponsor's product. When listeners submitted entries to one of *Vox Pop*'s many contests, for instance, they often took care to point out their loyalty to the program's sponsor—as if that would win them special favor. Mrs. Leo Lewalski of Michigan City, Indiana, let the program's announcers know that, even though she was a woman, she used the sponsor's shaving cream and that if she won the watch she sought, she would showcase the shaving cream's value. "Back again w/ another [submission]," she wrote. "Have been using Molle on my hands and arms—to keep them smooth—so as to show off that 'Norma' watch to best advantage. Hoping I have better luck this time." Some in *Mary Marlin*'s audience wielded their consumer clout less subtly. When listeners feared Proctor and Gamble would discontinue the program, many wrote the sponsor explaining that they loved the program—and Proctor and Gamble products. Many others went farther, though, and hinted—or occasionally threatened—they would boycott the sponsor's goods if the company cancelled the program. Their views should count, these listeners suggested, because they bought the advertised products; and because they were consumers some felt they had a means to make their views meaningful to centralized and distant corporate powers.[22]

Obviously, to the extent these listeners believed that as consumers they could influence those who controlled radio, a degree of self-delusion was involved. The voice listeners gained as consumers was a muffled one. Companies had little incentive to listen to individual, unorganized letter writers, no matter how much they claimed to love a product. Moreover, even when the volume of mail on a subject was sufficient to guide a sponsor's decisions, listeners did not effect systemic change. By locating their authority in their purchasing power, listeners accepted the new rules of consumerism. Corporations held much of society's cultural and political clout, and individuals participated in the power dynamic as buyers, not as citizens with inherent authority.

But that does not mean listeners' efforts to nudge radio were not important. They were. It is important to see how people have worked within the order they inhabited, not just how they have opposed it. In the 1930s radio's listeners often used their identity as consumers to say to an impersonal system controlled by large corporations, "I matter; listen to me." Certainly that was an extremely limited power, but it could give listeners a sense that they might count in the enormous public arenas of their mass society. Even as radio helped consolidate cultural authority in the hands of commercial powers, listeners used the medium's consumerism to declare their own importance within that system.

Listeners may not have been fully successful here, but in the process they helped shape the meaning of the medium. Despite the centralized corporate control of broadcasting, radio was not simply a commercial vehicle to listeners.[23] Their acceptance of consumerism on the air clearly indicates the pervasiveness of consumer values by the 1930s and the power of centrally controlled broadcasting. At the same time, listener responses reveal the possibility of haggling over the meanings of that medium. Even here, there was some room within the system for listeners to find their own meanings. And they did so.

"When You Can't Find a Friend, You've Still Got the Radio"

When Grace Squires visited her daughter in Washington, D.C., in 1938, her daughter made sure they saw all the important sites. Or almost all of them, Squires thought. After finishing her tour, Squires felt they had missed the most significant spot of all. "Do take me to Alexandria, so I can see where Mary Marlin lives," she asked her daughter. Mary Marlin was the wife of a senator, and later in the 1930s, she was a senator herself. At least that was how she appeared on the radio soap opera *The Story of Mary Marlin*. She was, in fact, a fictional character. "But mother," Squires's daughter replied, "that is a radio story."

"Well of course I knew it," admitted Squires in a letter she wrote to Mary Marlin, "but you are so real to me. For I have wept with you Mary Marlin, have been hurt with you, have caught your courage, and have loved your loved ones." For Squires, Mary Marlin was not simply a fictional character. She was a friend who visited Squires in her home every day. Mary Marlin helped Squires manage the connections between her private sphere and a world beyond Batavia, in upstate New York: Squires tied her daily life into Mary Marlin's circle and used that knot to help her personally. Real or not, Squires had come to love Mary Marlin and her friends, had shared their experiences and suffering, and had come to rely on them for guidance and inspiration. Through the radio, Squires felt she developed new relationships—intimate bonds that enabled her to enter into a new community, a community united by the air.[24]

Squires was just one of the many listeners who used the radio to create a new sense of personal attachments and community resources. In the meanings they found in radio, listeners revealed their sense that the wider world increasingly pressed upon their daily experiences, eroding their efficacy even in their own lives and washing away the support traditional communities might provide. At least as much as anything else, broadcasting intensified this trend, connecting listeners' lives with far-off events, ideas, and people.

But listeners did not see radio undermining their communities or their autonomy. Just the reverse. As listeners tuned into radio in the Depression, the medium enabled them to fashion their newly extended horizons into familiar patterns. Listeners developed intimate relationships with the voices they heard on the air, relationships that offered listeners the resources they sought. They personalized the seemingly impersonal arena of mass culture to allow them to gain a sense of control in their own expanding worlds.

The voices that listeners heard on the air, of course, spoke to an enormous and anonymous audience, not differentiated individuals. Yet in the air, listeners discovered not simply sound, but fellowship, sources of specialized expertise, wisdom, guidance, and personal connections. Listeners would find radio valuable, in part then, because the medium seemed to reinvigorate—even as it altered—their sources of support and community. Of course, radio-fostered communities could not match the face-to-face social exchange that might take place in traditional arenas. Indeed, we might call these radio communities ethereal ones—and not just because the air through which radio waves flew was commonly known as ether in the early years of broadcasting.[25] And yet, to those latching onto the new communal forms, they proved invaluable.

Radio felt so crucial, in part, because to many by the 1930s, the communities they traditionally relied upon felt very much in flux. As a mass society—with its vast lattices interconnecting the nation—infringed upon ordinary American's experiences, they felt their control of their own lives slipping. In that modern world, they could no longer find adequate guidance or connections to authority in their traditional personal relationships and communities. Over time, developments from the railroads to chain stores to movies to automobiles undercut the independence of local communities. With the social and economic crisis of the Depression, many individuals' senses of local autonomy came crashing down. When sociologists Robert and Helen Lynd visited Muncie, Indiana in the mid-1930s, they saw a town with its locality-based communities decaying—and unsure what would replace them. The Depression, they found, had jolted Americans, awakening them "from a sense of being at home in a familiar world to the shock of living as an atom in a universe dangerously too big."[26]

Radio played a fundamental role in the process of welding individuals to a national society, of pushing a realm that extended far beyond listeners' own spheres of influence into their lives. The new medium dramatically transformed its listeners' senses of their worlds' boundaries. An array of social scientists studying radio in the 1930s observed that radio's most critical impact was its tendency to dissolve geographic divides and to connect listeners to the world beyond their own physical space. Interviewing rural families in

Nebraska and Illinois, for instance, William Robinson of the Office of Radio Research at Columbia University found that having a radio boosted one's interest in national and international affairs. Robert and Helen Lynd noted the same phenomenon: listening to a national radio network station "carries people away from localism and gives them direct access to the more popular stereotypes in the national life," they wrote in 1937. This might be, however, a disorienting process. While tuning individuals into a broader world, the Lynds and others argued, the new medium ruptured a listener's sense of local connection and control, possibly turning him or her into a "lost individual, untied in any active sense to community-wide life and values."[27]

Listeners certainly found radio expanding the scope of their worlds, but they did not experience that in the disquieting terms the Lynds feared. Because listeners were able to take the public communications between centralized voices and a mass audience and reinterpret them as intimate, private relationships, listeners, in fact, often used the medium to tie themselves actively to a new community life, one that offered a new measure of individual control. They conceived of an abstract social environment in concrete terms, experiencing the voices on the air as friends and integrating those friends into their daily lives.

Often when news analyst H. V. Kaltenborn delivered one of his radio commentaries, Geo. Burglehaus, a listener from Sumner, Washington, wished he could reply in person. He satisfied himself by writing Kaltenborn a long letter, outlining his own views on the news. He closed by inviting the commentator to drop in for a visit; after all, Burglehaus reasoned, Kaltenborn was a familiar friend.[28] Burglehaus had never met Kaltenborn, but he still assumed that the famous radio analyst would be interested in reading his opinions, in getting to know him personally. Hearing Kaltenborn on the air, Burglehaus, like many other listeners, imagined a genuine rapport linked him and the commentator.

This tendency to see the broad world that radio presented in terms of personal relationships was particularly evident in listener responses to a program such as *Vox Pop*. *Vox Pop* first hit the national network airwaves in 1935 as a program featuring spontaneous sidewalk interviews. It rapidly evolved into a quiz show in which its announcers broadcast from public spaces and asked members of their audience a mixture of trivia, logic, and simply bizarre questions. Through the late 1930s the program seemed almost designed to foster a sense of community among listeners: the announcers constantly strove to involve listeners as well as their audience on the streets. At times the announcers all but turned the airwaves into an introduction to their listeners' attics and neighborhoods. The announcers called on their listening audience to send in questions for use on the air. They held scavenger hunts, asking listeners to find objects, from covered bridges to wooden cigar-store Indians to funny hats, and send in descriptions or photographs. They held look-alike contests, again calling for photos. They asked listeners to send them interesting facts about their hometowns, and discussed them on the air. Even the questions the announcers asked contestants were designed to spark listener participation. According to Parks Johnson, the chief creative force behind the program as well as its lead announcer, the ideal *Vox Pop* questions gave listeners something to discuss at home and write the program about.[29]

Such efforts to encourage listeners to feel personally connected to particular radio programs and their broadcasters were common. By the 1930s, programs often created environments that made it possible for listeners to feel personally included. Radio helped popularize a new style of music, for instance: the soft, intimate singing called crooning prized vocal sincerity over projection and range. Talk show host Mary Margaret McBride frequently read parts of listener letters on the air, emphasizing her audience's role in the program. Many programs included formal incentives to write in. Those for children, for instance, often called on listeners to join formally in radio clubs. In 1935 nearly a million children joined Orphan Annie's Secret Society, receiving membership certification and the like. Such radio clubs urged children to feel a link between the radio program, themselves, and others who joined the special listening family. Chicago station WLS, "The Prairie Farmer Station," used the idea of family overtly as it suggested listeners see themselves as part of an ethereal neighborhood. Each year in the 1930s WLS distributed the *WLS Family Album*, introducing the station's performers and administrators to listeners. Consciously inviting listeners to become members of the WLS family, the albums included photographs, nicknames, and personal details about selected listeners as well as stars. Many broadcasters worked hard to relate to their audience in individual terms. That does not mean, though, that those broadcasters actually created intimate relationships with listeners; they simply provided a fertile climate in which those bonds could grow. It ultimately took listener actions to develop their feelings of connection for themselves.[30]

Listeners did that, forging relationships with the voices on the air to degrees that sometimes shocked broadcasters. A moderately popular program such as *Vox Pop*—which typically ranked only slightly above average in listenership ratings—drew about a thousand letters a week in the late 1930s.[31] Through such letters, many listeners actively interacted with the programs they heard. To *Vox Pop*, listeners sent in potential questions; they disputed the answers the announcers gave; they told stories about themselves and their homes; they asked questions of all sorts as though the announcers were reference librarians; they answered questions the announcers asked.

In such interaction with the program, listeners saw themselves participating in personal relationships of sorts with the announcers. When Johnson and his coannouncer Wally Butterworth wondered what a geoduck was several times on the program in late 1937, responses from listeners poured in for months. G. E. Hauson of Tacoma, Washington, was one of many who offered to take Johnson and Butterworth digging for the large clams. He had an excellent chef for a neighbor, who would be happy to cook their harvest, he added. The Vox Poppers, Hauson seemed to suggest, were a welcome part of his neighborhood. Seattle's Harold Peters did Hauson one better, shipping a geoduck right to New York. To Hauson, Peters, and other listeners, Johnson and Butterworth were not impersonal announcers, but friends. Listeners frequently referred to listening to Johnson and Butterworth as a visit from the announcers. *Vox Pop* was not simply entertainment to Ethel Strintz of Elkhart, Indiana. When the announcers told a wondering woman on the air that all pigs had curly tails, Strintz sent in photographic evidence to the contrary to set her friends straight. "We don't know what we would do without our radio," she concluded her letter, "it is so much company."[32] Not entertainment, but company.

Listeners who tuned into other sorts of programs also frequently incorporated the world they heard on the air into such private relationships. *Vox Pop* simply offers a detailed example of a common listener response. Many of H. V. Kaltenborn's listeners, for instance, wrote him to offer their opinions on the news, to ask him to explain things to them. They felt the same personal connection to Kaltenborn that Geo. Burglehaus did. "I feel as if you were a friend tried and true I have heard your voice so often in our own home," wrote Gertrude Christine of Louisville. "If you ever come to Louisville I hope to at least see you." Kaltenborn lived far away, but many listeners conceived of him as a neighbor, not as a distant ambassador from the wider world. "We have come to regard you as a friend and neighbor and you know how us mortals like to talk with our friends and neighbors," Charles and Elizabeth Dunlap of suburban Philadelphia wrote, explaining why they felt comfortable telling the commentator their views. Similarly, some of *The Story of Mary Marlin*'s listeners felt a personal affinity for its creator and writer, Jane Crusinberry: they sent her Christmas cards, even long after the program left the air. Along with letters, Mary Margaret McBride received photographs, recipes, suggested remedies when she was sick, and various gifts from her listeners. And when popular radio comedian Eddie Cantor revealed his shirt and sock sizes on his fortieth birthday, listeners sent him fifteen thousand birthday presents. Many listeners collected pictures of their favorite radio stars. In St. Francois, Missouri, Mrs. Dalton pasted pictures of musicians above her radio set; that way, she said, when she heard a performer on the air, it seemed as though she were listening to the musician in person.[33]

Crafting such radio friendships, listeners often did not bother distinguishing the radio voices of real people from those played by actors reading a script. Some listeners believed fictional characters they heard were real. Some, like the *Mary Marlin* fan Grace Squires, knew otherwise but did not see a need to build a rigid divide between what on the radio was real and what was not. Fans of the most popular of the daytime dramatic radio forms, the soap opera, often responded to the programs as though they depicted actual people and events. Based on her interviews with one hundred soap opera fans in the New York City area, Herta Herzog of Columbia's Office of Radio Research concluded, "The listeners studied do not experience the sketches as fictitious or imaginary. They take them as reality and listen to them in terms of their own personal problems." Not surprisingly, then, many of the women Herzog interviewed did not understand questions about their opinions of the actresses and actors playing various characters. Herzog may have overstated the case, but a sizable minority of the fans of *The Story of Mary Marlin* who wrote to Mary Marlin accepted that program as real. Some, like Jessie Penn of Milwaukee, who warned Mary Marlin against a bad marriage, wrote with advice for the character. Beulah deRocher of Kalamazoo, Michigan, wondered if she could get a job as governess for one of the characters, a talented musician. Probably not, she concluded: "I am about as musical as a donkey." More often, though, *Mary Marlin*'s fans knew the program was fictitious, but believed in it to some degree anyway. Howard Matteson of Fall River, Massachusetts, wrote to the program complaining about a change in the actresses playing the title character, but he also sent in the name of a doctor he felt could cure the blindness of one character. And Alice Neale, who began her letter, "Of course it is only a story," concluded with the fervent wish to ease the suffering of Mary's husband.[34]

The implications were palpable: even an imaginary character might be a friend, and a fictional plot might feel like a part of a listener's world. "You are not giving us a fairy story," wrote Catherine White of Milford, New Hampshire. "You are giving us Life." Or, to reorient White's point, in soap operas, listeners often found people and experiences they integrated into their own lives. Listeners tended to favor the particular soaps that most directly resonated in their own experiences. Mrs. Lefkowitz, a New York City Jew like the title characters of her favorite program, dubbed *The Goldbergs* her favorite serial. In it, she said, "I see my own and my mother's life." The Goldbergs did not simply reflect Lefkowitz's life from afar, though. They were, Lefkowitz continued, old, familiar friends who dropped in for a visit. Many of Mary Marlin's fans also considered her an integral person in their lives. Grace

Roggenstein of Utica, New York arranged her morning routine around her daily visitor: "My whole schedule was planned for my favorite program, and I hustled around every day to get my morning work done, so at 10:30 I could sit and relax, have a cigarette and enjoy this interesting story. . . . I am devoted to these friends of mine and really felt they were a part of my life." In Herzog's broad study of soap opera listeners she found such thinking common. Listeners, Herzog observed, "do not want to lose the story characters they have grown to consider as belonging to their family."[35]

Soap opera fans were not the only listeners to imagine intimate relationships with fictitious characters. Listeners created a sense of community based on evening serial dramas and adventure programs too. Some listeners believed that the characters in the exceedingly popular serial *Amos 'n' Andy* were real in a fashion. After listening to an episode in which Amos and Andy struggled to run their taxi company, Judge Joseph Mulgreen reduced the length of Harlem taxi driver Charles Davis's assault sentence. Based on Amos and Andy's experiences, Mulgreen said, he concluded "that taxi life is not a sinecure." For some listeners, the relationship with the program's characters felt concrete enough to inspire a vague sense of responsibility to help their radio friends. When Amos and Andy mentioned that they lacked a typewriter for their business, 1,880 listeners sent machines to them care of NBC. One listener even sent in a dog after Amos lost his. (In this case, though, many white listeners probably felt personally connected to the fictional world of *Amos 'n' Andy* precisely because of its unreality: many of those listeners who sent Amos and Andy office supplies probably would have refused face-to-face friendships with blacks.)[36] Children were particularly prone to see radio characters as real and as a part of the listener's world. In interviews, several children said that they knew that characters in books and comic strips were fictitious, but those they heard on the radio seemed genuine: "You can *hear* the real ones on the air," explained an eight-year-old boy.[37] That some listeners participated in ethereal communities that included imaginary people may not be so surprising: on the air Mary Marlin could sound as close to a listener's life as H. V. Kaltenborn.

In the end, listeners could imagine concrete personal relationships through the air because many listeners believed radio enabled them to get to know a performer directly: the mass medium, many felt, conveyed not only sound, but authentic personality. *Vox Pop*'s announcers, Johnson and Butterworth, not only did their job well, but were "good fellows" and "fine gentlemen," according to listeners in San Diego and Hartford. More than an insightful commentator, Kaltenborn was "brave" as well as "modest and gracious" in the ears of several Lincoln, Nebraska, and New York City listeners. One listener went so far as to suggest she knew Kaltenborn's innermost feelings.

"You, you that I know so well, even though, I have never met you, I know and understand how you must feel, very deeply in your heart," claimed Concha Marin of New York City. In laboratory experiments in the early 1930s the psychologists Hadley Cantril and Gordon Allport found this phenomenon as well: radio listeners associated certain personalities with certain voices. That Cantril and Allport concluded that those associations often turned out to be accurate no doubt would have strengthened listeners' vague sense that radio could foster a new version of the meaningful private relationships they sought.[38]

These ethereal relationships were meaningful to many listeners. Listeners came to draw upon them for advice and support in living their own lives. This was crucial. The ways listeners actually used these relationships in their daily lives reveal just how valuable the arrival of broadcasting proved to be to popular audiences. Listeners discovered sociability and a sense of belonging in their new radio friendships; despite the seemingly one-way nature of radio communication, listeners felt reciprocal exchanges did take place. The listeners who grasped onto these misty relationships were not irrational nuts, but people who sought the resources of a community. Listeners could and did rely on their new webs of personal connections for specialized information and expertise, and for wisdom, life guidance, and help confronting the Depression and the modern world.

Most obviously, radio communities filled a social gap. The medium, listeners claimed, helped them combat loneliness by bringing them in touch with new friends. One Brooklyn woman considered her radio the most essential piece of furniture in her house. Her husband was in a sanitarium with tuberculosis and her radio got her through the lonesome months, she said. A survey in the Northeast found that for 83 percent of listeners, the radio made them less lonely. In Manhattan, a mother believed that radio helped her daughter, an only child, pass the time when she was alone. "It takes the place of companionship," the mother said, "and proves very valuable in this respect." The Manhattan mom did not mean to suggest that radio kept her daughter from seeing other people, but that radio provided an activity that eased the daughter's social yearnings when she could not see other people. To the extent that listening to the radio diminished actual face-to-face interaction and replaced it with the long-distance contact of listening, imagining, and occasionally writing to programs, the medium certainly shrank social connection. But, as many listeners in the 1930s saw it, radio replaced isolation, not interactions, and provided them with voices they understood as friends. Listeners repeatedly referred to Mary Marlin as a dear or an old friend. And Ray of Seattle, who apparently felt he was on a first-name basis with the *Vox Pop* announcers, offered to catch Johnson and Butterworth several salmon, put them on ice, and ship them to New York. Advertising executive Harold Johnson marveled at the intimacy listeners expressed in the hordes of "chatty, friendly" letters written to the programs his agency produced.[39]

Listeners could understand radio personalities as friends because they often perceived broadcasting as a part of a two-way conversation. Each year millions of listeners envisioned themselves engaging in discussions with radio personalities through the airwaves and the mail—and many times the radio personalities wrote back. Kaltenborn's listeners eagerly jumped at the chance to share their stories and views with the commentator. "I was so pleased to hear you say last night (via radio) that you would like to hear from your radio audience," Lily Sykes of Maplewood, New Jersey, wrote in 1930 when the rules for audience-speaker interaction were still being made plain. "I have been waiting to hear that for years." Similarly, *Vox Pop*'s listeners frequently debated comments made on the air, elaborated on them, or told personal stories they inspired. When, prodded by one listener's boast, *Vox Pop* announced that Keene, New Hampshire, had the widest paved street in the country, listeners from Iowa to New Jersey leapt in to contest the claim. And after *Vox Pop* sent listeners on a hunt for covered bridges, Pinto Colvig of Hollywood was convinced that he had an interested audience for his reminiscences about the covered bridges of his youth. "The BEST use, however, that I made of it was a swell place for us kids to hide from the far seeing eyes of Gran'ma when we went there for a hide-out for to snitch a few puffs from the forbidden cigarette," he wrote. Herta Herzog overdramatized the gulf between neighbors in her study of soap opera listeners, but in her observations about the very human interaction listeners felt they shared with radio personalities and characters, she uncovered a common sentiment. "In a way the radio seems to have taken the place of the neighbor," she wrote. "The neighbor as a competitor has become the stranger, while the radio in its aloofness is the thing humanly near the listener. It offers friends who are 'wonderful and kind.'"[40]

These ethereal friends were particularly special, though, because they had access to kinds of expertise that listeners lacked. Listeners in the 1930s relished the chance to bond with friends who seemed comparatively well-connected and able to explain the world and how to live in it. At times listeners turned to their radio communities for specific information, for instance. *Vox Pop* listeners, for instance, presumed that the program's announcers should be able to answer any nagging question, and sent in gaggles of queries. William Vogel of Pittsburgh wanted to know why the dots in a polka-dot handkerchief wore out before the rest of the fabric. Stunned at how quickly her hair grew right after she cut it, Emily Pierce of Philadelphia wondered if women found that to be typical. Both turned to their expanded community.

And when Hollywood songwriter Pinto Colvig could not find a word to rhyme with "wolf" he too sought out the "smart ones" on the air.[41] Through the airwaves, listeners found they had some degree of access to friends who were experts of all kinds. Listeners with questions about current events frequently wrote to Kaltenborn seeking explanations. Betty Chalk of New York City contacted the commentator for help on a school paper about president Franklin Roosevelt; George Hull sought Kaltenborn's opinion on a book; Inez Pelins of Bayonne, New Jersey, asked her expert friend to explain the Muscle Shoals bill in Congress to her. In the rural Midwest, women often sent letters to the host of the station WLS's *Homemakers' Hour*, Martha Crane, for professional advice on household problems; in the summer of 1929 alone, 20,511 wrote seeking Crane's help with their canning.[42]

Many listeners also believed their newly expanded circle of personal resources included wise friends and role models, people who could give listeners welcome advice on how to live their lives and manage their private problems. Soap opera fans, for instance, often saw their favorite shows providing them with examples of how to behave in daily life. Esther Norman of Leavenworth, Kansas, praised Mary Marlin for offering listeners a model of poise and gentleness. And in New York City, a widow and *Stella Dallas* fan moved out of her son's crowded apartment when she heard her radio heroine sacrifice her own comfort for her daughter's. Their radio friends, many listeners believed, had valuable life lessons to impart. *The Story of Mary Marlin* gave Catherine White of Milford, New Hampshire, her greatest inspirations and taught her the nature of suffering, she claimed. A Mrs. Ryan of New York City similarly lauded her favorite radio serial: "If I had listened to it when I was young it would have helped me avoid a lot of pit-falls," she said. The practice of turning to a soap opera for advice was quite typical, according to Herta Herzog's study of listeners. Many listeners, Herzog found, had no one they could rely on for guidance: the women Herzog interviewed reported that they could not talk to their husbands about problems in the family, and only 20 percent said they saw friends they could turn to for help. One-third of listeners explicitly said they listened to soap operas as a source of interpersonal advice. "I really don't have anybody to talk to, and I would have needed advice in the tragedy which happened to my daughter," said one listener. "Dr. Brent [a character in *The Road of Life*] is such a fine man. It helps me to listen to him."[43]

All sorts of listeners, not just soap opera fans, drew on their ethereal communities for this sort of intimate guidance and to help them place themselves in the wider world around them. Children, for instance, often looked to radio programs for examples they might follow as they matured. One twelve-year-old girl listed *The Green Hornet* as her favorite show because it featured a secretary, and she hoped that by learning what secretaries did, she could work as one in the future. "You learn about life from the radio stories," said another twelve-year-old girl. "The stories are like life themselves, and so you can learn how it is when you grow up." And Kaltenborn's listeners wrote to him not only because they valued his news expertise, but because they respected his wisdom and judgment. Doris Hough, a high school student in Braintree, Massachusetts, wrote to Kaltenborn for advice about what career she should pursue.[44] Adults too sought Kaltenborn's guidance and help with their careers.

Occasionally, listeners turned to their ethereal communities for concrete support. In 1933, frustrated by his business failings, C. P. Lohse of Bogota, New Jersey, decided that the ideal of autonomous individuals in control of their own lives no longer worked: the world was too complex and power too centralized for ordinary people to succeed relying only upon their own resources. And, he might have added, he found that the relationships individuals could forge within local communities were not effective in a vast world. So he wrote Kaltenborn asking for help making business connections: "I have come to the conclusion that in these very serious times it is only possible to find employment through the good offices and with the aid of a friend of the Court. 'Rugged individualism' alone will not do it. It is for this reason and in consideration of your wide circle of friends that I write you."[45] Lohse, like others who sought Kaltenborn's aid finding a job, counted on their imagined webs of connections to help them manage the complex society of twentieth-century America.

Throughout this process, few, if any, listeners saw themselves engaged in a deliberate sleight of hand, consciously exchanging one form of communal links for another. That is, it is wrong to suggest that ethereal communities simply undercut more familiar ones; the reality here was complex. The public mass communications ushered in a novel array of personal relationships with disembodied voices, but listeners did not experience this as an attack on older interpersonal ties. In fact, in some respects, the collective experience of listening to the radio might cement traditional bonds. Part of the power of the new medium lay in buttressing existing modes of communication in order to build new relationships among various listeners. At the same time, though, radio did not always reinforce face-to-face relationships either. It was possible that, in some circumstances, mass communication might replace personal communication, that imagined experiences and relationships might edge out actual lived ones.

Listening to the radio periodically allowed audiences to enter into communities with other listeners as well as ones with the voices on the air. This might happen, as several historians have observed, when listeners found common ground with and imagined links to other distant listeners, people who may not have seen each other, but who shared both the experience of listening to the same programs and the cultural touchstones those programs disseminated. Historian Lizabeth Cohen asserts, for instance, that radio did more than any other medium to create a shared working-class culture that extended beyond the bounds of traditional neighborhoods and communities. The medium, in other words, made it easier for listeners to imagine sharing some experiences with listeners they might never see. Contemporary students of radio at times noted this phenomenon too. "The listener has an imaginative sense of participation in a common activity," psychologists Hadley Cantril and Gordon Allport observed in 1935. "He knows that others are listening with him and in this way feels a community of interest with people outside his home." Listeners in the Depression might have agreed. A 1935 study of unemployed Americans found that many poor families felt that radio alone allowed them access to the culture shared through their society. "It keeps us in touch with the outside world," one unemployed father said.[46]

More often, though, communities of listeners grew up in private settings. The act of listening itself could draw listeners together physically and create a common, face-to-face experience among them. Listeners reified the value of their private spheres by listening in the home to the public discussions on the air and bringing those discussions into settings such as circles of family and friends. Since radio's listeners commonly imported the public sphere into the private one, it is perhaps unsurprising that, as we have already seen, they also blurred the lines in reverse—by taking their expectations for the private sphere and imposing them on the public arena they heard through the airwaves.

Radio, some listeners believed in the 1930s, could serve as a kind of personal social glue, potentially joining together families, friends, and perhaps neighbors. In the long run, it is questionable whether radio tightened such private communal ties, but listeners did feel at times that they and those they knew well could come together around the sounds produced for mass consumption. After all, simply listening to the radio was often a shared experience: Americans in the 1930s typically listened with groups of family or friends. As a regular Monday night activity in 1937 Bill Oliver of Covina, California, gathered with friends to listen to *Vox Pop* and to answer the quiz questions. And Julia Bingham's Business Women's Club in New York City held discussion groups twice a week to listen to Kaltenborn's commentaries and explore the issues he raised. In some circumstances, radio might pull families together too. According to one study of unemployed families, children from families with radios spent more time at home than children whose families lacked access to the medium.[47]

In Seattle, Andrew McQuaker and his family not only listened to *Vox Pop* together regularly, but they engaged in friendly family feuds over the answers to the program's questions afterward as well. For the McQuakers and many others, listening did more than bring them together physically: radio programs offered groups of listeners a mutual reference. During the Munich crisis in 1938, listeners from New York to Texas reported greeting each other with, "Did you hear Kaltenborn today?" Kaltenborn's coverage of events in Europe gave neighbors and associates a shared experience from the public sphere through which they could relate to each other individually. More mundane broadcasts could also provide listeners with common cultural ties. A majority of parents and children, according to one study, believed radio programs offered families grounds for discussion and social connections. Moreover, 88 percent of New York City children surveyed in the mid-1930s said they frequently discussed programs with their friends. Children often played games with their peers based on radio adventures. After losing the family radio, Jean, a nine-year-old who had frequently played Lone Ranger, worried that her friends would laugh at her for being unable to follow the story. To some degree, then, the common cultural experiences mass communication encouraged could translate into face-to-face ties.[48] Individuals could use mass-produced programs and their abstract sense of shared culture to modify and reinforce more personal and concrete bonds.

But the medium did not inherently have the effect. Radio in the Depression was often a collective experience, but that does not mean it always improved the connections within that collective. For all the social listening that took place, in the 1930s the overwhelming majority of listeners did not prefer to listen in groups: in one study 35 percent of listeners said they enjoyed radio more when they listened alone and 54 percent said they had no preference. And by the 1940s, as the cost of the technology dropped, listening to radio was becoming a more solitary activity. More tellingly, in his study of rural listening, William Robinson of Columbia University's Office of Radio Research found that radio neither strengthened nor fractured families of listeners, but simply exaggerated existing tendencies. In some families, parents and children came together to discuss programs; in others, radio wedged members apart. Vicarious experiences could provide fodder for interpersonal bonds—or replacements for them. When 63 percent of fourth-to-eighth graders in a study reported listening to the radio during family dinners, we might, and many in the 1930s did, interpret that as a way in which families could spend time together and establish common ground. But, as researcher Herta Herzog noted, we could also read that as a threat to family interaction, as listening took the place of actual exchange.[49]

That appropriately raises the question, how valid were the relationships audiences began widely crafting in the Depression—not just the ones between family members, but those that joined listeners to ethereal voices? Certainly listeners felt they benefited from the radio communities they established. But equally certainly, the imagined interactions of listening and occasional letter writing could not provide the same level of personal exchange and potential support that face-to-face relationships might: radio permitted only a moderate degree of social participation. For all that listeners might consider Mary Marlin or H. V. Kaltenborn a friend, the relationship was largely unrequited: neither the character nor the commentator even knew most of their listeners. Ultimately, the communities of the air were rooted in nothing more concrete than the air itself. The new socially constructed communities of the twentieth century were stitched together with thin, often temporary threads.

And yet, they were ties that mattered to those who felt them. Radio communities were not as deeply interwoven as traditional ones, but they did meet some important listener needs. Like many of Kaltenborn's listeners during the harrowing days of the Munich crisis, Agnes Thompson of Lawrence, Kansas, believed that the news analyst had given her the support, the insight, and the wisdom to make sense of and survive in a world that, like it or not, in fact extended far beyond the limits of her physical community: "I hardly know how I could have withstood the anxiety without it," she wrote.[50] Soap opera fans did not generally escape into the imaginary worlds of the serials, but used those serials to help them explain their own worlds. These listeners were not foolishly seduced by the magic of the new medium. They came to rely on new communal forms, because those were ones that met listeners' modern needs.

Listening to voices broadcast to much of the nation, we might well say that listeners entered into a public sphere. Radio created an arena in which ideas were shared across a range of impersonally connected people; listeners participated in a common civil society and culture that existed on the air.[51] But even as radio consequently radically enlarged a public sphere, making it difficult to avoid, listeners entered that sphere in very private ways. Ostensibly, a listener's relationship with a speaker on the mass medium of radio was a public one. Yet by creating personal ties to those speakers, listeners related to radio's world through intimate friends; listeners transformed the large and abstract world of radio into a private one. In the process, they blurred the line between the public and the private, repainting a disconcertingly massive public sphere as more manageable and familiar private ones.

We may—indeed we should—criticize the radio-fostered connections as inadequate and not truly interactive. But the new relationships many listeners adopted were not unreasonable adaptation efforts. Traditional face-to-face

communities could not meet all one's needs in the mass society of the twentieth century, and Americans needed an alternative. As radio helped extend
listeners' worlds, those listeners often cast their new horizons in comfortable shapes. Radio seemed to many listeners to offer a reasonable chance to
construct meaningful communities. By re-envisioning the growing mass culture in terms of their private interactions, listeners gained a sense of efficacy
within it.

To public intellectuals like William Orton and James Rorty, the medium that
would help shape the century's communication exacerbated the dislocations
of their rising mass culture. It challenged individuals' autonomy in their own
lives and their ability to speak out to the broader world. But many Americans
also found in radio ways to gain a place for themselves within that same culture. By using radio to personalize the public world that touched their lives,
they felt they could claim a measure of importance and control. The practice
of creating and relying upon ethereal relationships made a real difference
for both individual listeners and the modern United States. As we will see in
the two chapters that follow, those seemingly personal connections mattered
not just for listeners in their daily lives, but in terms of the political life of
the nation as well. By using the medium both to craft new styles of political
communication and to make impersonal political relationships feel intimate,
radio listeners and politicians restructured the meaning of democracy for a
mass society. As listeners relied upon the relationships they forged of the air,
many felt they could participate in intimate but national political communities in which they mattered. Others believed the personal links they forged
connected them to champions they could count on for help. Janet Bonthuis,
remember, saw *Vox Pop*'s Parks Johnson standing up for her when she could
not help herself. Some listeners were more deliberate in seeking out speakers
who promised to make listeners' causes their own. These various resources
were not equally valuable, but they all represented listeners' efforts to find
ways to count as individuals even as they felt their lives enfolded into their
growing mass society. In the meanings audiences found in radio's popular
programming—in ethereal communities, even in accepting a consumer identity—listeners felt they could personalize and gain traction in their world.

    Like most of those listeners, Grace Squires, the ardent fan who sought
out Mary Marlin's home on a trip to Washington, D.C., did not explain her
fondness for radio in such abstract terms. Each day as she listened to the
program she worked on crocheting a bedspread, with each section representing a different episode. Finished, she did not consider the work a bedspread, but a scrapbook of the lives of people she had come to know and
love. The bedspread, then, told Squires an inspirational story. Mary Marlin's

life, Squires believed, served as a model that could help her better get along in the world. In a vague sense, Squires understood her relationship to that story: in her letter of devotion for Mary Marlin, Squires took care to gush about the sponsor's products as well.[52] But Squires did not mind her role: it was, in part, as a consumer that she felt she counted. More than that, she relished what she gained from the program. For Squires, *The Story of Mary Marlin* offered her a source of friendship and guidance. Many listeners wanted something quite similar. They created that new sense of personal connection and individual significance through the air.

3

RADIO'S DEMOCRACY

The Politics of the Fireside

Visiting Washington, D.C., in early 1935, university student William Nels made the rounds of the federal government buildings. In front of the Treasury Building the West Virginian gaped, in awe of its size and apparent distance from his life. "I couldn't help but think of its vastness and I wondered whether we citizens of this nation really realized the significance of our government machinery," he later mused. "We often wonder whether all of the work that goes on in these buildings really means much or what it should to the individual American. It seems so aloof and one cannot imagine himself as a part of the government." As he left the capital, it seemed to Nels that he could have little connection to what took place inside those massive buildings. Shortly after his return home, though, Nels sat in front of his radio set and heard a warm, clear voice reach out from Washington and call him a friend. He heard president Franklin Roosevelt explain in simple, plain language what the federal government was doing and why. The government that had seemed so far away as he walked amongst its castles, suddenly felt meaningful in his own house. Roosevelt's radio talk awoke in Nels a sense of the personal significance of the central government, and a sense of his own importance within it.[1]

For Nels and millions of other listeners, radio made possible a new relationship with their government. Through the medium, listeners found the sense of being vitally connected to a distant, abstract, and centralized government. Listening to political broadcasts in the 1930s, Americans like William Nels began to feel personal ties linking them to their officials; radio made it possible for listeners to re-envision the impersonal relationship between politician and citizen in twentieth-century America as an intimate one. And for many of those same listeners, the mass communication form also enabled them to learn easily and directly about their nation's political life, and to find a sense of place within it. The ethereal communities many Americans participated in as they listened to popular programming had political manifestations as well then. Americans in the Depression decade discovered the chance to enter a new national political community and sphere of discussion, one bound by imagined personal bonds and by shared public information. Radio, many felt, might help them overcome the challenge of making the individual relevant in the twentieth century's enormous political arena. To most listeners, this welcome development would reinvigorate democracy in the modern era. In the process, though, radio's listeners and politicians redefined the idea and practice of democracy for broadcasting's century.

The new radio democracy relied on mass communication and listeners' abilities to personalize it in order to adapt broadly democratic principles to fit a mass society. In its most general sense, democracy may be said to represent a system in which the public can participate effectively in the meaningful discussion of civic issues. By the 1930s, many Americans had come to feel that centralized decision-making and the enormous public sphere overwhelmed individual voices and made such efficacy hard to come by. In radio, though, they believed they had found a way to change that. By imagining personal relationships and a sense of community with central officials, listeners used radio to craft private modes of interaction through which they could enter the public sphere. In other words, using radio, listeners gained the sense that they mattered in their national political life.

Doing so, though, changed the nature of that political life and participation within it. Mass democracy further empowered central authorities. More than that, radio politics meant that democratic participation could become a private act, one that often amounted to spectatorship. Listeners could feel publicly involved simply by listening at their metaphorical firesides, not by speaking out themselves or by communicating with other citizens. To the majority of listeners in the 1930s who feared that their modern society threatened individuals' chances to enter into civic life at all, this new participation came as a relief. Radio's democracy offered those listeners a way to count in the world and in public discussion. At the same time, though, it changed—in some ominous ways—what it meant to count and where those discussions took place.

For many Americans there was a real need for change of some kind by the 1930s. Though not all listeners phrased it as such, the early twentieth century posed a real challenge to democracy: by the start of the Depression, many of radio's listeners felt detached from meaningful political authority or discussion. How could Americans effectively engage in politics when the decisions that seemed to matter most increasingly took place far away? How could ordinary citizens take part in public discussion in a mass society where the public consisted of millions and only mass-produced voices could reach such far-flung ears? By almost any standard, popular participation in politics faded after the collapse of the Populist movement in the late 1890s. Voter turnout for presidential elections, for instance, fell from 77 percent in the last decades of the 1800s to just 52 percent in the 1920s. Big businesses gained increasing influence over national politics. At the same time, the decades-long trend toward centralization of economic and political power reduced the autonomy of local communities and made local civic involvement less meaningful. Moreover, Americans saw their venues to participate in political life wane. As reform efforts weakened political parties, American politics lost what historian Michael McGerr calls a "vital democratic theater." With parties sponsoring rallies and other communal affairs, political involvement in the 1800s was a shared event; the decline of such visible political activity partially removed the discussion of public matters from public space.[2] By the crisis of the Depression, Americans had to consider whether and how democratic values applied in an age of national, mass politics.

Using radio to personalize the political realm, though, listeners again found they could comfortably function as individuals within their mass culture. As Americans came to understand their leading medium of mass communication, it served as a vehicle that might recreate a sense of intimacy within the public sphere, enabling ordinary people to feel they could have a meaningful but new civic role. Drawing upon radio broadcasts, many listeners essentially imagined that they, elected officials and, through those officials, at times a national public, belonged to a shared and familiar political community. Long before the days of television, broadcasting became America's "democratic theater."[3]

The men and women who listened to political broadcasting played a crucial role in determining its meanings, interpreting the broadcasts for themselves and endorsing with their votes politicians whose radio voices sang in tune with those interpretations. Those listeners, though, were not alone in placing radio on a democratic stage. The politicians of the 1930s who spoke on the medium and who sought to define its political meanings conceptualized radio—or, often, fought over it—in similar terms. As listeners began envisioning their radio democracy, no individual played a more important role in helping to structure it than America's first radio president, Franklin Roosevelt. Roosevelt, more than any single politician, pioneered a new politics of the air. He modeled a style of radio address that helped make seemingly personal ties between audiences and politicians possible. He took advantage of radio's ability to reach large audiences and to distribute public information widely. Together these two elements became the cornerstones that listeners used to build a sense of intimate connection to their government and public affairs. Americans, listening to radio politicians such as Roosevelt, came to feel having a voice in civic discussion meant being informed about the issues and sharing radio-woven bonds with public officials.

Not all Americans, of course, applauded the new politics of the airwaves; radio's democracy had its share of skeptics in the 1930s. Critics scattered across the political spectrum occasionally worried that the emotional intimacy of radio might enable skilled politicians to replace democracy with manipulation. With the charismatic Roosevelt dominating the radio dial, Republicans in particular voiced worries the microphone might give rise to a fascist-style dictatorship in the United States. Radio did, in fact, enhance the authority of the executive branch, helping to cast the president as the embodiment of the will of a national populace. And—as we will see in chapter 4 as well—talented speakers did find in the medium a means of securing greater influence.

None of that, however, went as far as the most alarmed critics feared. In fact, the biggest risk broadcasting posed in the United States came from the more subtle ways that it contorted the meanings of democracy. Through their mass communication, Americans changed the nature of political participation, making it into a more private and passive matter. If radio offered Americans a new "democratic theater," it was one located in individual homes, not in shared public spaces. A listener could engage civic matters without ever engaging citizens. Involvement could mean little more than listening to distant authorities speak. Listeners could imagine themselves active citizens because they were informed, rather than because they took public stands. Participating in a mass democracy was often a vicarious experience. Radio by itself did not offer most listeners an effective means of influencing political agendas. The mass medium put popular opinion more squarely in the political arena, but ideas and priorities in the radio democracy largely flowed one way—originating from a central authority.

But very few Americans in the 1930s focused on the political function of radio with a critical eye. Radio's political flaws did not register with millions of listeners because they were so grateful for what the medium offered them. For the many, many listeners like William Nels, who suddenly felt a part of their national government for the first time, radio offered a chance to connect to an otherwise distant and incomprehensible power. That newfound participation felt genuine; something real was gained. We would be silly simply to condemn radio's democracy for stifling participation. Many Americans already felt they had no ways as individuals to participate in a distant political system that included an enormous nation; radio changed that. In a mass society, can we conduct national discussions without mass media? But the weaknesses present in the modern democracy developing in the Depression are still plain. As political discussion became increasingly a private activity shaped and initiated by those at the center, something real was lost. Broadcasting helped bring listeners into a political community with their national government, but we should question the quality of that ethereal connection. At the same time though, we should take note that—and consider why—listeners, as mass-media politics developed, found that connection so valuable.

Roosevelt on the Radio

For William Nels, this newfound sense of his place within the nation's political life came about as he listened to president Franklin Roosevelt on the air. That was hardly coincidental. All over the country, men and women alike helped form radio's political meanings listening to Roosevelt. Many politicians took to the airwaves in the Depression decade, but none had the reach or impact that Roosevelt did. Radio, in important respects, made Roosevelt; and he, in important respects, made radio. Roosevelt benefited enormously from a medium that was coming of age in the early 1930s. He, in turn, helped transform radio's political role, popularizing the personal conversations that would be dubbed fireside chats as well as overseeing the expansion of other governmental and political broadcasting—all helping to make radio an important vehicle for political discussion. Roosevelt's radio presence, therefore, provided crucial raw materials out of which many listeners constructed broadcasting's political meanings for the 1930s and beyond.

As the millions of Americans who listened, and the hundreds of other politicians who took to the air, made sense of radio in the arena of politics, Roosevelt's understanding of radio resonated. The president believed broadcasting could serve twin aims for him: it could widely distribute information to create an informed general citizenry; and it could enable Roosevelt to enhance his political power. A politically educated populace and a popularly elected executive with the power to govern effectively, Roosevelt might have said, enhanced democracy; certainly, he argued, radio served democratic aims in the United States. Roosevelt's dual radio goals, however, also suggested that the line between informing people and garnering political power was blurry, and that broadcasting would make the distinction between mass democracy and mass manipulation hazier still. Not surprisingly, the president's use of radio drew critics. But in the 1930s his vision proved more influential than controversial: rival politicians generally sought to follow the radio president's lead even as they critiqued his position. Roosevelt's sense of the medium helped reshape radio not because his New Deal reformed the existing broadcasting structure; it did not. But in his distinctive use of that structure, Roosevelt offered a model other politicians adopted in varying forms. And, more critically, that model became one many listeners built upon as they interpreted radio's place in the public life of the nation.

Listeners had, of course, heard politics on the air long before Roosevelt delivered the first fireside chat in 1933, but the medium found its identity as a political forum in the Depression decade. President Woodrow Wilson had spoken on the air months before regularly scheduled radio broadcasts debuted with the coverage of the 1920 election returns by Pittsburgh's KDKA. And presidents Warren Harding, Calvin Coolidge, and Herbert Hoover also took to the microphone—with increasing frequency. Similarly, candidates for office increasingly relied upon radio. The Democratic Party's election campaign spending on radio increased sixteen-fold from 1924 to 1928, a trend that slowed but certainly continued in the 1930s: from the 1932 campaign to that of 1936 the Democratic Party's radio expenditures jumped from $340,000 to $840,000. And throughout that time period, the Republican Party routinely topped those outlays. But if the need to use radio politically was clear by the end of the 1920s, just how to do so and for what ends were murkier. Many politicians tried unsuccessfully to take their formal political speeches to the air. Although Hoover, for instance, recognized the medium's growing importance—he appeared on national radio twenty-seven times in 1930—he was a notoriously bad radio speaker. The *Nation* speculated that Hoover turned millions against him simply by refusing to limit his addresses to his allotted time and running into popular shows with his ponderous arguments.[4] Just as broadcasting in general settled into its mature form in the early 1930s, politically radio only fully arrived in those years as well. In terms of distribution of receivers and serious popular acceptance, radio was primed for a major political role by the start of the Depression decade.

As governor of New York in the late 1920s, Roosevelt had frequently spoken on the radio. In Albany he had showcased the simple, intimate style of radio address that CBS Washington bureau chief Harry Butcher later dubbed a fireside chat. Still, the nation could not have anticipated the impact Roosevelt would have when he sat down before the microphones March 12, 1933, and began, "My friends, I want to talk to the people of the United States for a few minutes about banking. . . ." Listeners were enthralled. Roosevelt had abandoned the flamboyant style of the stump speech in front of a huge crowd that some politicians used on the air; instead he employed an intimate approach. "The President likes to think of the audience as being a few people around his fireside," said Roosevelt's press secretary, Stephen Early. In plain, common language Roosevelt explained an issue, what the government was doing, and why. Not all of Roosevelt's subsequent addresses were dubbed chats; he gave twenty national broadcasts in 1933, and only four of them were considered fireside chats.[5] But in his addresses he had helped develop a new political role for radio. It could enable a leader to bring particular issues, information, and arguments to the people. It could create a personal connection between leaders and listeners. Roosevelt's plans to reorganize the banking system in 1933 required popular acceptance, for instance, and he used that first radio chat to help secure that consent. Roosevelt had taken radio and turned it into an integral part of his political arsenal.

This was entirely intentional. The shape of Roosevelt's radio addresses reflected the two inseparable aims shared by the president and the man responsible for his media relations, press secretary Early: to distribute information openly and to enhance their political strength. Early, a former journalist, and Roosevelt both believed that it was the government's responsibility to disseminate as much news as possible. In a drastic change from Hoover's day, Roosevelt opened his doors to twice-weekly press conferences. "I want to be a preaching President," he said. "Like my cousin." Early himself considered installing an extreme open-door policy for the press: "I may even take the door off my office so anybody can walk in at any time," he joked.[6] This vision of the government as an educational news agency helped mold Roosevelt's radio style, fostering addresses designed to explain and inform.

Explaining and informing could, of course, also enable a skilled speaker to marshal political clout. And the pursuit of such authority also guided Roosevelt and Early. Bluntly, that meant finding ways to use radio to win over as large an audience as possible: popularity—with the public, not just among politicians—would become a crucial form of power. Consequently, Early and Roosevelt sought to turn the president's radio appearances into major events. Roosevelt learned from Hoover's overexposure and broadcast less frequently and for shorter periods than his predecessor. And despite listener requests, he kept his fireside chats in reserve, pulling them out only for the rare occasion when he felt the situation demanded it. On occasion he sought out subjects to discuss that would warrant a fireside chat, but he held off speaking without cause. Following Early's guidance, Roosevelt guarded his radio voice closely, allowing no imitators, building listener anticipation. Then when the president did speak on the air, he and Early worked to turn the fireside chats themselves into big news. Several weeks before an address Early let the press know one was coming. Mounds of publicity followed as newspapers speculated on the subject of the upcoming talk. And Early took care to avoid needlessly upsetting listeners: he scheduled the president's chats to run after the popular evening shows had finished so no listener would resent missing out on a favorite.[7]

Roosevelt's concern with his radio popularity paid off. The president became a radio celebrity. Listener letters and telegrams, from a wide range of writers, swamped the White House after each fireside chat: he received almost four times as much mail per capita as the previously most written-to president, Woodrow Wilson, during World War I. In Moultrie, Georgia, the town often put radio loudspeakers in the courthouse square when Roosevelt spoke, and area farmers, most of whom did not own radios, packed the square to hear him. Radio gossip writer Helen Nelson gave Roosevelt's fireside chats the same four-ace ranking she reserved for the likes of comic stars Ed Wynn and Eddie Cantor. And the president must have been pleased by the listener rating surveys he requested. Roosevelt typically drew a listenership that would have made any sponsor drool. A 1939 poll found that 38 percent of Americans claimed they listened to Roosevelt's major radio addresses. And for highly significant chats, listenership could run much higher: 79 percent of Americans heard the president's address following the bombing of Pearl Harbor.[8] This sort of radio popularity mattered in Washington as well as Radio City: it bolstered the president's claim to speak with the public's voice.

Although Roosevelt's chats deservedly garner most of the attention, the president found additional ways to use radio to serve his intertwined aims. He limited his own radio appearances, but encouraged various members of the administration to appear on the air frequently to explain what the government was doing. Each week, for instance, a cabinet official spoke on NBC's program *Planned Recovery*. Various government departments also collaborated with networks to produce informative entertainment. By 1940 some forty-two federal agencies had taken to the microphone to varying degrees. And from 1938 to 1940, with wary enthusiasm from Roosevelt, the Interior Department had its own studio in which it produced programs to discuss government activities. Roosevelt also made the most of radio as a campaign tool, speaking as both a candidate—and buying time to do so—and as the president—whose addresses were aired by stations for their news content.[9]

All of this, not surprisingly, angered Roosevelt's opponents. Republicans contended, quite legitimately, that all the radio appearances by Roosevelt and other members of his administration served the president's political ends. They protested the New Dealers' extraordinary access to the airwaves. Because the president appointed the board that licensed radio stations, some Republicans complained, few stations would risk offending Roosevelt. "The radio, controlled by the Administration through its licensing power, was made the spokesman of the New Deal," the *New York Herald Tribune* cried in 1934. By the 1936 election Republican grumbling about Roosevelt's use of the medium grew louder. The party, for instance, launched a magazine, *Uncensored*, with a section devoted to refuting points made by New Dealers on the air. Even as the Roosevelt administration expanded the government's involvement in radio programming in the late 1930s, scattered critics had not been silenced. Representative Everett Dirksen of Illinois in 1940 derided one federal agency's broadcasting unit as "nothing but a political bureau." In general, such complaints had some merit: the networks frequently tried to ingratiate themselves with the administration.[10] Republicans had reason to equate Roosevelt's radio democracy with propaganda politics.

The president's opponents, however, did not always stand firm against using radio for political gain. Many Republicans simply wanted access to the same propaganda machinery. Michigan senator Arthur Vandenberg believed that Roosevelt's dominance of the air stifled criticism of the president. In the 1936 campaign, though, he tried to beat Roosevelt at the master's game. Vandenberg arranged to "debate" the president on CBS: the senator squared off against recorded Roosevelt addresses Vandenberg edited together for his own purposes. Citing their no-recordings policies, many stations cut away from the debate—which, despite Vandenberg's tactics, inspired a new round of Republican complaints about Roosevelt's control of radio for his own ends.[11] Opponents had little sympathy for Roosevelt's position that he was simply carrying out his duty in a democracy, to inform the citizenry. Thanks in part to radio, the line between information and manipulation was increasingly fuzzy. But Republicans were equally eager to try to tar Roosevelt for crossing it and to follow his lead.

Drawing colossal radio audiences and imitators across the spectrum, Roosevelt helped to redefine political radio. Overwhelmingly he endorsed the existing broadcasting structure that had developed under Herbert Hoover's guidance; Roosevelt declined to use government oversight of radio to tinker with the system's commercial networks, and accepted its faith in capital-controlled radio as a source of free discussion.[12] But his use of that system helped transform political discourse in the United States. Communicating the news was, for Roosevelt, a source of power. Radio, more than any other medium, allowed the president to control the process of distributing news, to inform and influence the public directly. That enabled him to preserve his own democratic ideals at the same time that he won tremendous popular approval. And he did not hesitate to turn that popularity into pressure on Congress to pass his agenda. As Early put it, radio was invaluable for Roosevelt, and the president had created a wholly new relationship between the government and the medium.[13] It was a relationship that various other politicians—including local officials and Republicans such as New York City mayor Fiorello LaGuardia—imitated through the 1930s. And it was a relationship that benefited radio as well. The diverse politicians who innovated in broadcasting helped give radio a serious political role in America.

Consequently, when men and women turned on their radios in the Depression decade, they found a medium occasionally concerned with the administration of their state and society—and formed their own ideas about the political character of radio and modern democracy.

Radio Democracy: The Politics of Intimacy

Listening to Roosevelt on the air, many Americans came, like the president, to see radio as a vital piece of their political life—as a medium that could enhance, indeed preserve, democracy in the modern United States. Listeners and various radio politicians understood democracy as a political form in which the will of the people governed society. But even such a vague system seemed, to many, threatened in the 1930s. As local public spheres, so vibrant in the nineteenth century, lost some of their vitality in the twentieth century, meaningful popular participation in politics became more difficult. By exacerbating the rise of centralized politics, the economic and social crisis of the Depression left many individuals feeling divorced from the political process.

But radio, listeners and Roosevelt believed, could offer a solution. Radio, they felt, could help individual people to participate in the vast and impersonal modern public sphere. Radio, they felt, could help reconnect those individuals with their government and politicians by creating personal and informational cords linking ordinary citizens and central authorities. Not everyone in the 1930s cheered these developments, and indeed, looking back we can see that founding political participation on such bonds transforms the idea of democracy. But to most listeners and to the radio politician they heard most prominently, political radio's bonds of intimacy and information, in tandem, helped revive popular government in the United States.

To listeners, radio could bring people closer to distant politicians and insure that their government embodied the popular will in part because they saw in broadcasting the possibility of sincere and personal communication. For those who listened to Roosevelt on the radio, the medium was foremost a means of forming an intimate relationship with their leader. Listeners found a heart-to-heart quality in his speech that left them feeling Roosevelt had actually stopped by for a visit. They applauded his plain, easy-to-understand language, words that told them they mattered.[14] Some criticized this new, personal, style of political communication, but to most listeners, it was an innovation that made a positive and real difference. Many reported that simply listening to the president speak made them feel better about their Depressions troubles—an oratorical Sudafed, of sorts, easing painful symptoms without curing the cold. More importantly, Roosevelt's style enabled Americans to imagine a crucial human bond between citizen and politician. That connection made the radio democracy of the 1930s and the rest of the century possible. Because listeners felt they shared a personal relationship with the officials running the country, listeners also could feel they were a meaningful part of the otherwise distant and abstract arena of national politics.

Listening to the president speak, often in their own homes, Americans reconstructed the abstract and distant public sphere of national politics in terms of comfortable and familiar private relationships. Listeners came to feel Roosevelt spoke directly to them, that they knew him personally. He was not a vague and far-away figure, but a visiting friend. Writing to Roosevelt after a fireside chat, listeners repeatedly delighted in the intimacy Roosevelt's voice inspired. "It is almost beyond belief that a President has a heart to heart talk with his people over the radio—something that has never here to fore happened," wrote Edward Deininger of Reading, Pennsylvania. "I heartily approve of your getting into personal touch with 'your people.'" Like so many listeners, F. B. Graham of Dubuque, Iowa thought of a Roosevelt broadcast as a chance to sit down with the president, not just to listen to him: "The President of the greatest Nation on earth honored every home with a personal visit last night. He came into our living room in a kindly neighborly way and in simple words explained the great things he had done. . . . When his voice died away we realized our 'friend' had gone home again."[15]

This ethereal intimacy was not accidental. The public often responded favorably to speakers who mastered it, and Roosevelt and other radio politicians believed in its potential and did their best to nurture it. Instead of communicating with listeners only through public meetings or an intermediary such as print, with radio politicians could reach individual listeners directly. As he sat down to deliver a radio chat, Roosevelt consciously imagined himself speaking to an individual, not an audience of millions or even the thousands he might have addressed at a rally. He tried to envision specific people listening to him: "I tried to picture a mason at work on a new building, a girl behind a counter, a farmer in his field," he said. Roosevelt's secretary of labor, Frances Perkins, remembered that the president often appeared to forget that he was not actually sitting with a listener. "His head would nod and his hands would move in simple, natural, comfortable gestures," she wrote. "His face would smile and light up as though he were actually sitting on the front porch or in the parlor with them."[16]

Radio could bring ordinary people into political discussion, and, if used properly, the president believed, radio offered a chance to forge a connection with those people. Roosevelt and his speechwriters prepared his radio chats with typical listeners in mind: the president's aides kept his addresses short so he could speak slowly, and they relied heavily upon simple, common language. CBS vice president Henry Bellows and others believed that Hoover had spoken too formally on the air; the value of Roosevelt's ability to speak familiarly on the radio and convey his personality in his voice could not be overstated, Bellows suggested. Roosevelt, quipped humorist Will Rogers, "showed these radio announcers and our public speakers what to do with a vocabulary—leave it right in the dictionary where it belongs."[17] By speaking in plain terms, Roosevelt successfully conveyed his ideas to a wide range of Americans. And, probably more importantly, he let a wide range of Americans know he wanted them to understand his ideas.

That attention mattered to listeners. It drew them to the speaker and made the often-abstract issues he discussed real. Radio could be the vehicle that turned complex political matters into problems that made sense to all listeners. Many who wrote Roosevelt following his opening fireside chat, for instance, said that for the first time they understood the banking crisis that tortured the country in early 1933. Listener after listener thanked Roosevelt for explaining the nation's problems in terms that made sense even to those without education, cosmopolitan experience, or a dictionary. Roosevelt's "simple, calm and direct manner" had made him "understood at the cross roads of America," wrote one listener; another applauded the president's use of commonplace analogies such as baseball games. "The greatest compliment I have heard of the Presdts talk Sunday," wrote J. C. Cassell of rural Pennsylvania, "was by a child here who said—'the Presdts speech could not have ammounted to much because even I could understand it.'"[18]

Roosevelt was hardly the only politician seeking to connect personally with listeners through the air. Radio politicians of the 1930s did not all attempt to clone Roosevelt's style, but many did strive to build a sense of intimacy with their listeners. Louisiana senator Huey Long, after Roosevelt the nation's leading radio politician of the early 1930s, spoke rapidly on the air, but his reliance on colloquial phrases helped build his reputation as a spokesman for the common folk. Occasionally crude, always informal, Long spoke to his listeners in a familiar voice: he borrowed heavily from the Bible and tossed out expressions at home on the street corners of the small-town South. Long, like Roosevelt and many others, believed that radio's ability to reach the masses required a politician to make the effort to speak to them directly. New York mayor Fiorello LaGuardia, who pioneered radio's use in municipal administration, also adopted a simple, personal style on the air. "He understood that [radio] was not merely a medium for making speeches. It was, to him, a medium for talking personally with real people about real-life problems. He talked to people about their problems in their own terms," recalled Morris

Novik, who ran New York City's municipal station under LaGuardia. "It was as though his heart had a tongue that spoke to other hearts."[19]

This intimacy transformed many Americans' sense of their political system. As many of Roosevelt's listeners understood matters, prior to the 1930s America's public life was run by distant and mechanical powers. But, many listeners felt that, thanks to new and adroit uses of radio, politics had become a human activity, with actual people, not impersonal forces, holding sway. The New Deal era, of course, established new layers of centralized bureaucracy, but to listeners, radio cut through that and made even national politics a personal endeavor, one based on bonds between familiar individuals. Just hearing him on the air convinced listeners they knew Roosevelt's personality, not just his politics. To various listeners, the president was unselfish, determined, or manly. They rejoiced in Roosevelt's humanness—lauding the president simply for being "human" so often that a latter-day observer might wonder if HAL and not Hoover held office before Roosevelt. "It seems to me," wrote New York City's Frederic Drake after Roosevelt's first fireside chat, "that last evening I felt that the President of the United States was practically a neighbor of mine and now I am also sure that a man can be President and human at the same time." Like so many others, Roy Crawford of Evanston, Illinois exulted over discovering the humanity of a politician for the first time. Roosevelt's radio address was "like a dear friend dropped into our home for a few minutes chat," he crowed. "There is no 'Gabriel over the White House.' There is a human in it! Thank Providence."[20]

In his Gabriel reference, Crawford would have touched a nerve with critics of Roosevelt's radio style. Released just after Roosevelt took office, the film *Gabriel over the White House* portrayed in heroic terms a president who lifted the nation out of the crisis by assuming dictatorial powers. For some, the emotional intimacy of Roosevelt's radio addresses had a dangerous seductive quality. No matter what it felt like, the president had not visited his listeners' homes; he had not had a conversation with them, but spoken to them. Listening to a charming speaker over the air could come to stand in for real interaction; hearing an intimate voice could take the place of more active political involvement. Roosevelt's radio voice, writer John Dos Passos claimed, made listeners feel his concern, feel his friendship, feel his optimism. "No wonder they all go to bed happy," Dos Passos wrote. "But what about it when they wake up and find the wagecut, the bank foreclosing just the same, prices going up on groceries at the chain stores?" All Roosevelt's warmth did, Dos Passos lamented, was blind listeners to the desperate need for reforms.[21]

On the other side of the political spectrum, occasional conservatives leveled a similar criticism. "Mr. Roosevelt undertakes to charm away the sensible fears of the people," editorialized the anti-Roosevelt *Chicago Tribune* after one fireside chat. The editors worried Roosevelt's personality would effect what they saw as a radical agenda. Herbert Hoover, more strongly still, disapproved of the human, emotional quality of Roosevelt's radio addresses. Emotion, Hoover repeatedly maintained, had no business in politics: it clouded the free discussion and evaluation of facts; such appeals amounted to propaganda that ultimately threatened the foundations of democracy. Hoover needed only to point to Nazi Germany for a vibrant illustration of his fears. Hoover exaggerated the danger, but he was not entirely mistaken: radio, for instance, inserted the emotional qualities of national politicians' personalities squarely into political debate. Of course, few considered Hoover long on personality, and certainly, he never mastered the emotional intimacy of radio. Neither did Roosevelt's 1936 presidential opponent, Alf Landon. Landon and his campaign workers hinted that Landon's lack of radio skill indicated that he was an honest and regular man, while they sought to attack Roosevelt as a "radio crooner."[22] That effort, and the whole of Landon's campaign, failed spectacularly. No matter what Roosevelt's critics might say, no matter how valid their points, the majority of Americans cheered the personal and human contact they felt radio could add to politics.

That contact, in turn, cheered listeners. Dr. New Deal's radio medicine—his simple language, his intimate style, his human presence—helped relieve some of the psychic pain of the Depression. It comforted listeners and lent them courage. "Talk to us some more," wrote Henry Blagden of Saranac, New York. "It is good medicine and just what we need." Listener upon listener agreed. They fed on the confidence they heard in Roosevelt's voice. His calm cheerfulness proved infectious. "Never has the radio seemed to be so great a boon to mankind as it was Sunday evening during your message to the American people," claimed Raymond Blatchley of Indianapolis. "I thank you for the message. You have brought us hope and have transmitted to us a good measure of added courage to meet the issues." Regardless of the dire issues he might discuss, Roosevelt's listeners found his addresses uplifting. At the height of the Depression, struggling listeners wrote to Roosevelt thanking him for giving them faith in the future; as war loomed in Europe and Asia late in the 1930s, listeners counted on Roosevelt's radio voice to calm their fears. In 1939, Mildred Stillman listened to a Roosevelt radio chat while her young daughter lay upstairs in their New York City home. After the broadcast, Stillman went to check on her daughter and found her awake, waiting to hear what the president had said. "I told her," Stillman wrote, "and she snuggled down saying, 'I knew he would comfort us!'"[23]

The comfort and courage that listeners drew from their radios was significant. Listeners suggested they needed that spiritual nourishment in the

Depression. According to the occasionally critical *New York Times*, even when Roosevelt offered nothing but old and unworkable ideas, he still made "a great contribution to good cheer" by speaking and sharing his "stout heart and high spirit." And that optimism could have political and economic ramifications. Roosevelt's first fireside chat, delivered at the height of the banking crisis in March 1933, inspired Americans with enough faith in the reforming banking system to redeposit their money in the institutions. In those first few days after Roosevelt's radio address, the banking system was saved. Radio proved a critical tool in forging what historian William Leuchtenburg considered Roosevelt's greatest single contribution to the politics of the 1930s: infusing the people with hope and courage.[24]

It was, however, the creation of a human bond between listener and national politician that proved to be radio's most enduring contribution to the politics of the twentieth century. Because of their imagined intimate relationships with the president, listeners believed Roosevelt wanted to hear from them, that their personal views and experiences counted—even in the huge national public arena. For the first time, they mattered in the political life of the nation, many consequently concluded. Listeners who wrote to Roosevelt clearly saw radio not as a one-way transmission, but as the foundation of an essential two-way relationship with the politician. Eben Carey of Chicago had often wanted to write to the president in the early years of the Depression, but it felt inappropriate for him, an ordinary person, to contact the political elite, he said. But after hearing Roosevelt speak on the air, he changed that view: like many others, listening to the president on the radio made Carey comfortable with the idea of reaching out to Roosevelt himself. "The human appeal in your voice tonight was too much for me and this message is the result," he said in a letter to the president. Consequently Roosevelt inspired unmatched barrages of letters—from Democrats and Republicans alike. From Chambersburg, Pennsylvania, George Ludwig wrote the president about his children's efforts to find work. "Now then I suppose you would enjoy hearing a fireside chat from one of the common citizens of this Nation," Ludwig began his story.[25]

Roosevelt obviously lacked the time to read all the letters his radio talks inspired; in terms of speaker-audience interaction, his talks resembled lectures far more than chats. But listeners in the 1930s did not hear them that way at all. A fireside chat prodded Reese Farnell of Talladega, Alabama, to write to Roosevelt to ask a few questions. Farnell assured Roosevelt that anytime the president wanted to continue the dialogue, there would be a space by the fireside for him.[26] Roosevelt's intimate radio style had created a personal bond; it drew Farnell to the government and won his loyalty for Roosevelt. That proved critical in shaping a new interpretation of democracy.

Radio Democracy: The Politics of Information

At the same time radio connected a centralized government with far-flung individuals through mystic cords of intimacy, the medium also enabled listeners to link themselves to a national polity through bonds of civic information. Political radio, Americans found, not only could render the vast public sphere accessible through relationships that resembled those in one's private life, but it could make that complex national arena newly comprehensible. Listeners and radio politicians alike believed that broadcasting would make possible—for the first time—the mass distribution of public information. In the 1930s, the prospect of mass information held out tremendous promise for a democracy, the promise of giving ordinary citizens an understanding of their government and expanding world, and a sense that they had a meaningful place within it. At the same time, political engagement based first on being informed in private rather than on gathering and acting in public changed the idea of democratic participation, for good and ill.

Americans early in the Depression suffered not only from economic deprivation, but also from information deprivation; to many, the trouble they had in understanding the economic and political affairs that governed their lives shook their sense of control and divorced them from meaningful democratic participation. Through radio, though, listeners felt they could overcome the failures of existing media and learn about the public life of their mass society. The print media, some listeners and politicians like Roosevelt alleged, veiled rather than revealed. To Roosevelt and his supporters, radio circumvented that barrier: the radio president saw the medium's ability to inform a mass public as its most vital virtue. Hence, radio politicians on national and local stages designed government broadcasting with educating the public in mind. The mass distribution of public information, they argued, was essential in a modern democracy. Of course, one politician's educational material is another's propaganda. To critics of radio politicians, radio did not support democracy, but endangered it; those critics rued lost checks on biased speech. What they did not account for was what the newly accessible materials meant to listeners. Overwhelmingly, listeners cheered that, finally, they had access to information about their public life. They felt empowered by their knowledge. They felt a new sense of connection to their government and importance in it.

When self-described staunch Republican H. L. Boyer of Geneva, New York, wrote to Roosevelt in March 1933, it was not to complain about the Democrat. Instead, Boyer lamented that he was sick of the "bunk" that passed for political news prior to the new president's first fireside chat. Americans—particularly in light of the Depression—desperately needed to understand what was going on in the world beyond their personal experience, he said. But, he added, no one he knew comprehended national institutions and economics. Boyer was hardly alone. Many Americans in the early decades of the twentieth century felt cut off from adequate access to information about the expanding arena of public affairs. As listeners saw it, before Roosevelt took to the radio, their government had kept them in the dark. Outside of elections, one listener complained, he heard nothing from his public officials. Alternatively, John Dolan of Lowell, Massachusetts, suggested, what he heard from past officials was worse than nothing: "speeches, meaningless in themselves, . . . expressed solely to confuse and confound."[27]

The problem, though, could not be blamed entirely on closed-mouthed national leaders. The real issue was, as the significance and visibility of local politics and parties ebbed, what information source could connect a far-flung population to a central government? Existing media did not meet citizens' needs in a political arena that increasingly encompassed a vast nation. The problem—for listeners and radio politicians alike—was one of metaphorical distance; listeners sought a means to secure a direct connection with events on the biggest political stage. Newspapers and other information sources, they felt, were too mediated—by distance, by private interests and by biased editors—to provide the missing link to the center. Myra Gardner and her friends in Detroit wanted to understand current events, but they did not feel they had enough access to accurate news. "The papers print only what they see fit to print," she complained. Too often, listeners lamented, big business's interests ruled the broadsheet.

In radio, though, they hoped to find a medium that could create a direct line of communication: elected politicians could leap across distances and speak to the citizenry without mediation. In Avondale Estates, Georgia, getting reliable news about national affairs was a challenge, W. Hamilton Lee believed; only what he heard on the air was trustworthy. "I believe if you knew how little real information we get in this part of the country," Lee wrote to Roosevelt, "you would deem it advisable to talk more frequently over the radio." Listening to Roosevelt on the air, Harry Goldman of Los Angeles cheered the new scene he saw unfolding: "The day of the newspapers and other indirect communication is passing."[28]

Roosevelt also lauded that development. Like many of his listeners—indeed, more so—the president had little faith in newspapers and saw radio as a welcome alternative. Because of what he saw as the biased mediation of reporters and editors, Roosevelt did not trust print media to convey news accurately. Throughout his tenure in office, Roosevelt made his displeasure with newspapers plain. The president liked to claim that 85 percent of the press disliked him. While that figure was an overly high estimate, the majority of the nation's newspapers did oppose Roosevelt throughout the Depression. A financial elite, Roosevelt and his colleagues asserted, controlled the press, limiting its voice and keeping it from speaking for the common interests. Moreover, Roosevelt disdained the interpretive journalism he saw as on the rise. As national events became more complex, reporters frequently used experts to help them explain events to readers; in Roosevelt's view, this denied citizens the chance to make their own decisions from plain facts. Consequently, he believed the print media could not be counted on as an information source: it neither represented the popular will nor presented unfiltered news. This, the president maintained, left the public with no means of informed participation in politics.[29]

But, like many of his listeners, Roosevelt and his administration saw broadcasting as a way around this informational bottleneck. Radio, he said, could "restore direct contact between the masses and their chosen leaders." The medium could allow Roosevelt to interpret the facts for his listeners himself; it could allow his administration to elude an outside interest's information filter and build an informational link directly to the public. In the hands of Roosevelt, mused Harry Butcher, a CBS vice president and friend of Roosevelt's staff, "radio short cut the press and enabled the President to get his message direct to the people." It was with this goal in mind that in 1938 commerce secretary Daniel Roper suggested that the government needed to produce its own radio news report. "The time has arrived for the Government to have its own radio factual commentator, with the objective of giving the public a truthful presentation of plans and purposes," he reported to Roosevelt.[30]

Roper's proposal was not enacted, but it was radio's ability to accomplish those aims that particularly excited Roosevelt. Indeed, to diverse politicians, the political wonder of the new medium lay in its unprecedented potential to distribute national news throughout the country. Radio, as Roosevelt saw it, offered a means of drastically improving ordinary listeners' understanding of public affairs. "I need not tell you that in my opinion radio now is one of the most effective mediums for the dissemination of information," Roosevelt wrote to NBC president Merlin Aylesworth in 1933. "It can not misrepresent nor misquote. It is far reaching and simultaneous in releasing messages given it for transmission to the Nation." Although he would later significantly qualify his endorsement of political broadcasting, as president, Hoover had in fact voiced a similar view of radio's civic importance: "As a medium for the universal dissemination of ideas it is an important factor in the formulation of sound public opinion and the promotion of good citizenship," Hoover wrote. Hoover primarily objected not to the use of radio to inform the public, but to his rival's radio style. Roosevelt, no doubt, saw that style as a critical means of conveying his information successfully. One of a president's greatest tasks, Roosevelt said, was to teach the nation; radio could provide him with a model classroom.[31]

Starting with his fireside chats, Roosevelt and his administration often turned that ideal into action. Radio politicians certainly understood they could use radio as a tool to push their own ends, but simultaneously, many who took to the air sought to educate their audience about public affairs. In 1935, for instance, Roosevelt's commissioner of education, John Studebaker, created the Emergency Educational Radio Programs project, a series of programs designed to teach listeners about the federal government's services. Throughout the decade other federal offices worked with the networks to produce similar informative programming. The popular *National Farm and Home Hour*, for example, included commodity price reports designed to help farmers market their crops in a national economy; it was a joint presentation of the Agriculture Department and NBC. And in 1938 the Department of the Interior set up an in-house facility to produce radio programs for network broadcasting. In a little over a year, the facility had produced and aired fifty-one programs, prompting secretary of the interior Harold Ickes to assert that the department was living up to Roosevelt's mandate that the "Government shall have full right to place all facts in its possession before the public." As Ickes claimed, "No other medium has reached so many people at such small expense to the Government."[32]

Roosevelt cast his priorities and his use of radio as he did because he believed popular access to civic information was fundamental to a democracy. That meant that a mass democracy demanded mass communication. "Factual and accurate news made available to all of our people is a basic essential of democracy," the president wrote; without a practical understanding of public events, the people would lose faith in their system and become disaffected. In the extended society of the United States by the 1930s, Roosevelt believed, broadcasting was essential to foster the sort of national discussion necessary for creating the wise public opinion that made popular government work. "Although radio has made a general contribution to the cultural life of our people," Roosevelt wrote, "it is the maintenance of the open forum for friendly and open debate that gives the American system of broadcasting preeminence."[33]

This general sense of the importance of civic awareness in a democracy and radio's potentially vital role in distributing information resonated among radio politicians outside the White House and Washington. New York's radio-savvy mayor, Fiorello LaGuardia, believed that providing listeners with information to help them thrive in an increasingly complex world was radio's supreme calling. Consequently, the mayor ardently supported public broadcasting in the form of municipal station WNYC. Indeed, in the late 1930s

LaGuardia fought with Roosevelt's Federal Communications Commission to enhance the status of New York's public station. Under LaGuardia, WNYC disseminated information about civil service exams to make it easier for listeners to get jobs; it aired food prices and market reports to help consumers manage the growing confusion of the marketplace. By distributing public information, LaGuardia believed, radio could empower the populace. He claimed,

> The utility of any instrument of government in a democracy must be judged in terms of its usefulness to strengthen and to maintain the system in which democracy functions. I am, therefore, proud of Radio Station WNYC when it serves the important cause of good government, by making the householders of our community better acquainted with the instrumentalities of our federal, state and local governments. . . . The economic and social functions which these agencies and laws perform are often obscure to the man in the street who does not have a direct contact or interest in these matters. None the less, it is of the utmost importance that these regulatory laws and government agencies be fully understood by the citizen of a democracy.

After all, LaGuardia concluded, the ability to manage one's own world wisely depended on having some understanding of it. Radio, asserted LaGuardia's appointed head of WNYC, Morris Novik, should be first and foremost "in the democracy business." Because it made the mass dissemination of information possible, radio politicians argued broadcasting could help the New York mayor's favored business to thrive.[34]

Of course, a medium that enabled particular politicians to disseminate their own versions of the public information easily and widely could be used to manipulate, not empower, citizens. Critics of Roosevelt rightly charged that the president had put in place a vast publicity machine. And the control over information that radio made possible could translate into centralized political power—power Roosevelt's opponents found ominous. It was no coincidence that the politicians who touted radio's value as a means to support popular government were also ones who used the medium to their political advantage. In Roosevelt's hands, radio served the Democratic cause as surely as a democratic one. Indeed, while he still had ready access to the airwaves as president, even Herbert Hoover applauded radio for distributing civic education. Out of office, though, Hoover became one of the most vocal critics of political broadcasting and particularly Roosevelt's use of radio. He repeatedly attacked the New Deal as government by propaganda, not public opinion. Roosevelt, Hoover declared, employed emotions, not facts, and thousands of workers to manipulate Americans' thoughts. Radio, he continued, was the Democrats' leading vehicle. By the late 1930s, the propagandistic side of political broadcasting worried some in Congress enough to prompt attempts to dismantle the administration's programming agencies. Press secretary Stephen Early tried to disarm such accusations, explaining that Roosevelt's radio publicity was, at its core, educational. "It is simply acquainting people with what their government is doing, or trying to do, and why," he said in a 1934 interview, "giving them the facts and letting them draw their own conclusions."[35] But, despite Early's claim, the line between education and propaganda is never so clear and bright. Roosevelt's administration did not misuse its authority to the degree critics feared, but as those critics charged, by giving officials greater control over information, broadcasting did make that blurriness more dangerous.

The problem, from the point of view of Roosevelt's opponents, stemmed from the same point Roosevelt and listeners cheered: the direct link radio made possible between politicians and citizens. To Hoover, the old print media had not failed democracy; the new mass media instead threatened popular sovereignty. A democratic society, Hoover argued, required not more speech, but truthful speech. Gresham's Law might apply to public information too. Radio—by offering unmoderated access to the public—invited politicians to distort the truth, to pollute the air with propaganda, and eventually to choke off free speech itself, Hoover worried. The print media, he said, kept politicians in line, filtering out their falsehoods. The editors of the press, Hoover said in a 1937 speech, "maintain hourly battle against [propaganda]. . . . They have the job of discrimination between propaganda and real news, between untruth and truth." In the 1930s, broadcasting weakened that editorial check. Over time that would help change the relationship of the press to the government, making it harder for the press to challenge the government directly and independently.[36] More immediately though—and from Hoover's point of view, more dangerously—radio could open a channel through which politicians might flood listeners with falsehoods. The instantaneous communication did not allow opponents or impartial parties to check facts in advance, Hoover said. And once a point was in the air, successfully refuting it was difficult, particularly for those out of power. Roosevelt, Hoover repeatedly claimed, took advantage of this access to the utmost, twisting the truth and using ferocious emotion, not solid arguments, to undercut his opponents. To Hoover, this threatened the foundation of American liberties. Dictators, he said with ample reason in the late 1930s, climbed to power on the sort of false mass communication radio made possible. No doubt looking at Adolf Hitler's rise in Germany, Hoover tried to warn the United States: "Liberty dies of the water from her own well—free speech—poisoned by untruth."[37]

Hoover's concerns were not entirely unfounded. But he failed to appreciate the concerns of ordinary Americans: just how cut off they felt from their growing central government, just how inadequate they felt older forms of public communication had proven in their emerging mass society. Simply, Hoover had missed just how deeply listeners valued the access to public information they found in radio. When politicians like Roosevelt took to the air, listeners relished the addresses as rare and valuable opportunities to learn about their political world. Orville Brown of Phoenix wrote to Roosevelt after his first fireside chat, "Your talk last night is one of the most significant addresses I have ever listened to and represents a policy which should be perhaps the greatest advancement any government has ever made—telling the people what is being done and why, cannot help but be for better understanding and advancement. We have too long been kept in ignorance of what our executives and legislators do until after it is done and then the propaganda often gives us the wrong impressions." Through radio, listeners believed they heard public affairs unveiled around them. After fireside chats, listener after listener reported that for the first time he or she understood the particular issue Roosevelt discussed. "The way you explain your plans to the people is so completely in the spirit of our governmental ideal that it raises the standard of our civic life, and helps to clear up some of our social and civic illiteracy," marveled Laura Lunde of Chicago after Roosevelt took to the air in 1937 and detailed the agenda he demanded of Congress. Radio, as Roosevelt used it, could educate America politically, most listeners felt. And that made it—and those who used it—tremendously welcome. "You can little know nor hardly appreciate the consolation of having heard from our president in such a lucid and direct manner," John Dolan of Lowell, Massachusetts, wrote to Roosevelt.[38]

Roosevelt's addresses, Dolan believed, "struck a true chord in American democracy." Listeners in the Depression often worried about the state of popular government, but not for the same reasons Hoover did. To listeners, the public information they heard on the air gave them a way to connect to their government; it gave them a sense that individuals counted even in their vast nation. In San Francisco, R. J. Barlet listened to Roosevelt's first chat and felt a part of the political process. Your radio talk, he wrote to the president, "is certainly a thing which makes one feel at least 'in' on what is being attempted. Give us more information." To the south, in Riverside, California, Marvin Coontz heard the same address and applauded Roosevelt for implicitly suggesting everyone mattered in the public arena: "I wish to express my gratitude, as an average citizen, for considering us important enough to explain your actions to." And in Pasadena, Queening de Mena concluded that, "we, as individuals really count and are a part of our great nation."[39] To many, like Dolan, the mass distribution of political information "struck a true chord in American democracy."

The awareness of their government's activities that Roosevelt's radio chats provided gave many listeners a small sense of control: they could understand the vast external events that shaped their lives; with public affairs no longer shrouded in mystery, listeners felt involved in what took place. Oliver Chryers of St. Paul never understood the whys and hows of business and the government until Roosevelt explained them on the air. All his life, he wrote, he had occupied unimportant positions, but Roosevelt had made him feel important: he was a small part of a larger process that he could comprehend. "My position in life has never risen above that of an employee," he wrote after hearing a fireside chat, "yet I consider that I am one cog in the wheel of the business for which I work. And as such, expect to stand or fall, fighting for the principles which you so abely set forth in your talk." Roosevelt, Chryers said, had instilled in him a sense of public responsibility. Chryers understood the situation, and had ideas about what reforms he wanted. He had no illusions that he could guide the nation, but he felt there were small actions he could take.[40] Through Roosevelt's use of radio to inform, listeners gained the sense that they belonged to a national political community and had contributions to make to it. That helped rejuvenate democracy in America. And redefined it at the same time.

Once and Future Ideals?

For the many Americans who felt their traditional civic ideals embattled in the modern era, radio then offered a welcome relandscaping of their nation's political terrain. As many listeners saw it, political broadcasting enabled them to update and therefore preserve their ideals of democratic government in an age of mass politics. In a democracy, listeners believed, ordinary citizens had to feel a part of the governing bodies that mattered in their lives. As listeners understood democracy, that meant citizens needed to be able to participate in discussions of civil matters with the polity. In other words, a government of the public will was only possible if the public had the chance to confer on the relevant issues. By the time of the Depression, many Americans questioned whether those ideals remained viable in their mass society. How could one participate in a public conversation when the public sphere encompassed a vast nation and its center lay far away?

The answer, many came to believe in the Depression, was in the air. The intimacy and information listeners found in radio enabled those men and women to feel they belonged to a national political community, one in which they could discuss their nation's fate. Listeners felt in the know and hence a part of the political process; they felt national leaders cared about them personally and hence could represent their views. To many Americans, using the modern mass communication made possible by broadcasting could reinvigorate traditional political ideals.

If this was a case of going back to the future, though, the new scenario looked very different than the past ideal. Radio democracy might in some sense preserve older political values in the face of an emerging mass society, but it did so by drastically reshaping the practice and meaning of those values. The connection listeners came to feel with their central government was based largely on imagined quasi-personal ties and a broad understanding of civic matters. This meant changing the very idea of what participation in public affairs meant. In the mass democracy of the twentieth century, participating in the public sphere could take place in the privacy one's own living room, away from the public. Being informed, not simply taking visible action, came to represent a legitimate form of involvement in the governing process. Listeners might take part in discussions of the key political issues facing their nation, but they did so largely as an audience, not as direct contributors to the conversation: information typically flowed one way, rather than being open for common debate. Consequently, as listeners identified a national political community in which public deliberation took place, they related to that community largely through central leaders: again, from their own homes, listeners could feel they participated in a civic discussion by hearing the radio address of an official, not by talking to other citizens. Public citizenship in the radio democracy could mean private spectatorship.

Clearly, from our vantage point decades later, we must question just how "democratic" this emerging radio democracy was and would be. In the decades since the 1930s, we have seen that Americans have at particular moments used broadcasting as a springboard for tremendous public action. But broadcasting also enabled Americans to eschew such activity entirely without provoking a sense of crisis. Broadcasting helped create an environment in which listeners could easily watch, listen to, and cheer or boo the political process, all without taking an active role in setting the terms of debate. If democracy is a private process, it is damaged.

And yet, the listeners who welcomed broadcasting to the political arena did so, in large part, because they saw radio supporting their democratic ideals. This is not a simple story of democratic declension: public communication has always posed challenges for self-government. We cannot simply condemn the new politics, but must also understand the needs it addressed. Listeners understood that their increasingly vast and interconnected society created a mass polity and demanded distinct ways to hold discussions with that polity. Radio, listeners genuinely felt, offered individuals a meaningful place in a seemingly inaccessible public sphere.

To people all over the country in the 1930s, popular politics seemed newly possible because the national government no longer felt detached from their lives. Following one of Roosevelt's fireside chats, Margaret Moore of Normal, Illinois, wrote to the president: "You remember how Jesus went about teaching the people; He didn't do much preaching. He walked among them and spoke as one having authority. When you go among the people, they have the joy of seeing you and hearing you say just what the Federal Government can and will do." In comparing Roosevelt to Jesus, Moore made an unusual, though not unheard of, leap. In suggesting that giving a talk on the radio amounted to going among the people, however, Moore voiced the sentiments of thousands of others. It is virtually a historical truism that the Roosevelt administration transformed Americans' relationship with the national state. "For the first time for many Americans, the federal government became an institution that was directly experienced," writes historian William Leuchtenburg, "People felt an interest in affairs in Washington they never had before." The policies of the New Deal of course meant the federal government touched people's lives in more extensive ways than ever before. But it was not these policies alone that convinced many Americans that the central government mattered to them and vice versa.[41] The thousands upon thousands of Americans who wrote to their president in the 1930s were not simply responding to the New Deal; broadcasting more generally had taught them that a network of potential resources existed beyond their own personal experiences.

It was the radio, listeners like Moore asserted, that brought Roosevelt and the government to ordinary Americans. Before Roosevelt took to the airwaves, listeners felt excluded from the national public life. With the intimacy and information offered by the new medium, suddenly listeners considered themselves participants in their national government. "There is nothing which gives the average man the feeling that he is an integral part of that mysterious institution known as 'our government' as sitting in on, as you aptly said, a report coming directly from you," Anne Clapp of Duluth wrote to the president. "We all 'belonged' if you know what I mean." As they listened to Roosevelt explain public affairs, many Americans came to understand the government's role in their lives; they came to feel they too had a role to play in national civic life. Newton Fetter listened to one of Roosevelt's chats with sixteen college students in Cambridge, Massachusetts. "May I assure you it was an inspiration to those youngsters," Fetter wrote the president. "Washington has seemed a long way off to them and to many of us of the older generation, too. Some how, last night, we were made too feel that we are a part of the government and that we have some responsibility."[42]

That last point was a crucial one for listeners. Feeling close to the center of national politics alone was not enough for democracy; citizens also needed to be able to take part in that government. And radio, listeners repeatedly declared, made participation, not just connection, possible. James Dunn, a particularly articulate listener from Chicago, saw the seeds of America's democratic ideals in Roosevelt's radio addresses. At its roots, he said, democracy required popular participation in genuine discussions of public issues. In the twentieth century's mass society, then, there was little room for democracy to flower—little room, that is, until Roosevelt resurrected old-style political discussions through radio. Brilliantly capturing the sentiments of many of his fellow listeners, Dunn wrote to the president: "You talked as easily and informally as a neighbor who had just dropped in to visit the folks. . . . There was no more authority, mystery or pose about your talk than in the old time political arguments around the stove and cracker barrel of a country store or in the old fashioned wooden Indian city cigar store. Truly Mr. Roosevelt you revived the modes and manners of the primitive forums of American democracy. The old town meeting is now a nation's meeting." The essence of democracy, many felt, lay in Dunn's "primitive forums," communities marked by the free and meaningful exchange of ideas. A complex and interconnected society, stretching across social and geographic distances had done away with the primitive, but radio, many listeners believed, could revive the past and restore democratic forums. Roosevelt and radio, Dunn declared, would bring back the cooperative pioneer days of bees and barn raisings.[43] Broadcasting, listeners believed, would make civic dialogue and political community possible in a mass society.

Obviously, from a technical standpoint, American broadcasting offered only one-way communication. But listeners saw it differently, using radio as a starting point to expand the bounds of political discussion beyond the local. Roosevelt voiced his views and plans on the air, and listeners wrote in to offer support as well as suggestions and critiques. Just as many listeners conceived of fireside chats as visits from the president, many came to believe that through radio and letter writing they could actually take a small part in a national policy conversation. From Birmingham, Michigan, came suggestions for achieving full employment; from Enid, Oklahoma, a call for a three-day period of national fasting and prayer. Listeners all over the country responded to Roosevelt's radio addresses by offering their own views on how to best meet the challenges the president cited. Certainly some listeners appreciated the scope of the government bureaucracy and understood that the president would probably never see a particular letter. But even those letter writers felt they could be a part of the governing process by expressing their views. After laying out his suggestions for improving business-labor relations, James Cullen of San Francisco voiced the newfound sense that participation was possible in a bureaucratic system. "I realize, of course, that this offer will not come within fifty feet, maybe miles of The President," he wrote, "but I hope that it will come within the reach of some aide who can use it, or the ideas it may suggest, to help our President in his noble endeavor."[44]

Occasionally, Roosevelt directly fostered this sense of dialogue: during some radio addresses he encouraged listeners to write him with ideas for solving the nation's problems. Even without Roosevelt's specific urging, though, listeners flooded the White House with their views. Listening to political addresses seemingly aimed at them personally convinced many Americans that their voices counted. They believed radio fostered a genuine dialogue with their government because they believed their letters to the president could make a difference. George Allen of Richmond, Virginia, ardently opposed Roosevelt, but he too wrote the president to try to influence the system. The president, he suggested, represented America, and as such should consider any citizen's views. "I do not know whether you will ever see this letter or not," Allen wrote. "I sincerely hope so, for it is a personal expression of thought from one American to his President."[45] Such faith was essential to sustain a democratic ideal.

Significantly, listeners' sense of engaging in a dialogue with their government did not arrive full blown with the first fireside chat of 1933. It developed through the 1930s as political radio grew more familiar and helped rouse people's sense of democratic involvement and helped them adapt it to fit their circumstances. Roosevelt's early fireside chats drew few policy suggestions; moreover, listeners who did send in such commentary periodically included apologies for foisting their opinions on the president. Over time, though, listeners came to see the opportunity radio could provide them to engage their leaders as a democratic right of sorts. The tenor of listener correspondence changed significantly. By the mid-1930s, letter writers stated their views boldly, as though it were expected of them. Late in the Depression, listeners could speak of writing to express their opinions as their public responsibility in a vast society. With powerful forces massed against Roosevelt, Paul Stern of New York City argued in 1938 that progressive Americans should rally around their cause by writing the president. "I believe it to be the duty of every true liberal to inform the President that the great mass of the common people is with him 100%," Stern wrote.[46]

Clearly, to someone like Stern, the private act of writing to a politician was not exactly a solitary matter. As the listener James Dunn had phrased it, American democracy existed in "primitive forums," arenas in which ordinary people came together to discuss public matters. When Dunn suggested radio recreated the communal democracy of the town meeting, but on a national scale, the metaphor was neither incidental nor unusual. Indeed, secretary of labor Frances Perkins suggested that Roosevelt had the same image in mind: the president, she said, hoped to create "a modern substitute for the old town meeting." As an array of listeners understood it, ethereal political addresses helped transform the United States from a loosely connected federation into a national polity. After hearing Roosevelt speak, listeners often reported feeling patriotic and a part of a national whole. "I believe that such talks will greatly cement the common people into a closer relationship than ever before," asserted James Fisher of Robesonia, Pennsylvania. A nationally shared understanding of public issues, Oscar Dallia of Chicago claimed, helped people from across the country see their overlapping plight.[47]

It would be easy to overstate the case, of course. The New Deal in general helped to unify the United States as a political and social entity. And the common fate some listeners noted remained common only for some: listeners did not report political radio bridging racial divides, for instance. Nonetheless, the broad sense of community felt by many listeners was essential for their idea of a national democracy: without the belief that across their country people shared some of the same challenges and values, the idea of nationwide popular rule did not make sense. Roosevelt exaggerated the matter when he suggested that "as millions of American families gather each day about their receivers, we become neighbors in a new and true sense."[48] But the president's talks did encourage listeners to see events through national lenses and enabled those listeners to redefine the enormous national public sphere as an accessible community.

As listeners did so, the meanings and practices of their democratic ideals were transformed. Radio and Roosevelt did not return listeners to the days of "primitive" democracy as the nostalgic James Dunn had suggested; rather, broadcasting enabled listeners to adapt democracy for the mass society of the twentieth century. That process revised the very notion of participation in self-government, making public discussion into a private, centrally initiated process.

Listening to the radio, Americans found they could take part in their national politics. But that participation often took place in the privacy of a living room. By privileging both public information and its private consumption, radio encouraged listeners to reconceptualize what it meant to be politically engaged. Popular expressions of democracy in the twentieth century no longer required gathering with other citizens, political rallies or marches, or even voting. Ideas of democratic citizenship still allowed for such visible political actions, but in broadcasting's age, listeners could well conceive of themselves as political actors without taking such public actions. Listeners gained a sense of civic participation because they were well informed on political matters and because they felt a sense of connection with their national leaders. Simply by listening to the radio, listeners could see themselves as politically savvy and active. In fact, most listeners in the Depression may well have been closer to their national issues and officials than in any period in the past. But information could replace action; imagined private bonds could replace bonding with the public at large. The theater of democracy, so central to earlier ideas of popular politics, could be hidden from view. Moreover, once hidden, citizens lost one important spur to political engagement: each other. The documented decline of shared forms of civic engagement in the late twentieth century has its roots in ideas of democratic participation that radio helped spread. What radio pushed, television shoved.[49] In radio's wake, a government of the people, by the people and for the people could take place without the people interacting.

In radio's democracy, then, even participation could become a relatively passive act. Listeners lauded their newfound access to political conversations and entered into those discussions in record numbers. But they did so as just that: listeners. The speaker, not the citizenry or representatives they actually knew, initiated the terms of debate. With exceptions of course, the most action Americans usually took was privately expressing their views in a letter to an official. Following one of Roosevelt's fireside chats, Bryan Doble of Hinsdale, Illinois, wrote to the president; he felt the need to justify himself, explaining that because he was reared in "the atmosphere of New England townmeetings," he felt he had the right to express his opinions to anyone, even the president. But, Doble had to admit, he never expected Roosevelt to see his letter; the best he hoped for was that "it may be tabulated by some one of the bright young men who surround you and added to the ledgers wherein the account of public sentiment is, I hope, kept."[50] Radio democracy might ensure that individual opinion mattered, but largely as a mark on a tally sheet, as a figure beyond the decimal point in a public opinion poll. Doble's voice might count statistically, but largely to the extent that he was one of a mass of listeners responding to issues Roosevelt had placed on the table. Radio helped revive political conversation, but the call and response of the new medium differed significantly from the give and take of Dunn's cracker barrel.

Ultimately, then, radio in the 1930s offered not a primitive forum of discursive democracy, but a seat in the audience and something of an illusion. The sort of private pseudo-dialogue radio encouraged aptly suited a huge society in which individuals felt far from sources of control: it enabled listeners to feel they participated in a national conversation, often with powerful authorities. Listeners clearly described these politics in terms of personal and traditional democratic forms. But those metaphors were deceiving. Roosevelt's fireside chats neither took place around a shared fireside nor involved the two-way exchange of a chat. That so many listeners saw this sort of conversation in terms of town meetings and face-to-face public dialogues is revealing, though: it indicates both the power of the medium to subtly transform American political expectations and the depth of listeners' needs for some sort of meaningful political engagement. Searching for a voice in a realm of mass politics, listeners accepted mass communication as necessary and came to see it in personal terms.

Such changes in political communication strengthened central officials and particularly the president, but—and this is essential—many listeners accepted that reorientation of power because they newly saw it as democratic. Broad-casting helped make a stronger executive not just a reality, but an expression of popular government. In the national political community of radio democracy, a mass citizenry connected with each other largely through the center. Consequently, national officials could claim to be in the closest contact with the whole of the public; the voice of a far-flung community of citizens could be said to be heard most clearly in the center. To much, though certainly not all, of America then, empowering a central executive seemed to foster a government based on popular consent. Listening to Roosevelt's fireside chats, many listeners concluded that it was the president who most completely embodied the people's will. Through radio, listeners claimed, Roosevelt made his message accountable to all of America. The president thus represented the whole of the populace more purely than particular members of Congress who were biased by special interests and could speak for only small regions of the country, listeners asserted. Benjamin Weis of Columbus, Ohio, took such thinking farther than many listeners, but he was hardly alone when he suggested that the expression of popular sovereignty and hence power lay with the president, not the legislature. "You," Weis wrote to Roosevelt, "are the chosen captain and Congress is expected to enact legislation in accordance with your demands."[51]

This prospective shift in the political balance of power roused the fears of critics of Roosevelt and radio politics in the 1930s. To opponents of the president, once again, radio looked more like a foundation for dictatorship than a framework for democracy. In Europe, fascist leaders used the airwaves in their successful pursuits of individual authority; to Herbert Hoover and others, Roosevelt likewise did not represent the public will, but used radio to manipulate it to his own advantage. Such concerns could be applied to more local forums of radio politics too: some critics leveled similar charges against New York mayor Fiorello LaGuardia for his use of the city's radio station.[52]

In fact, opponents of successful radio politicians overstated their case a bit. In the United States, radio did not become a simple autocratic tool. The critics were right, however, that radio altered the nature of political authority in America. Roosevelt's supporters declared that the president embodied the clearest expression of the popular will; this suggested a shifting sense of where the focal point of a democracy lay: not in disparate and far-flung interests, but in a central authority that might bring them together. Moreover, a politician with the radio access and expertise of Roosevelt gained extraordinary influence over public opinion and political loyalties. And that popularity could translate into political clout—clout that shifted the balance of power toward the executive branch.

This did not spell the end of democracy as Hoover feared, but it did notably change its practice. Most listeners readily accepted increasing the authority of the president as a simple update to their older values. Democracy, the fluent James Dunn explained, depended on leaders who brought public debate to the people—something he saw Roosevelt doing with radio: "Dictators dictate, Mr. President, Democrats discuss."[53] Perhaps. But in suggesting that a national leader speaking to a mass audience at home embodied discussion and democracy, listeners offered a new idea of just what it meant to participate in a government of, by, and for the people.

Critics of Roosevelt's use of radio were not, of course, entirely wrong. Like so many other radio listeners, historian Arthur Schlesinger, Sr., wrote a letter to Roosevelt to express his admiration for the president's use of the medium. In the past, Schlesinger said, he had not been guided by party loyalty, but voted according to issues at hand; listening to Roosevelt on the air, though, convinced the historian that he and the president stood together. "I could see that you were *my* man and (don't overlook this) I resolved that I should be *your* man," he wrote Roosevelt in 1935.[54] As the chapter that follows discusses, radio enabled listeners to connect with public figures who seemed to represent them fully, individuals who had voices that could resonate in a mass society. And radio enabled those skilled speakers to build up tremendous personal followings, and to turn that popularity into sometimes worrisome power.

But to listener upon listener, political radio seemed not only to empower the speaker, but the audience as well. "Your early radio addresses from the White House sealed our friendship," Schlesinger gushed in his letter to the president. "You not only told me about your plans and hopes but also about some of your difficulties." As so many listeners did, the historian found the sense of intimacy and the mass distribution of civic information made possible by radio transformative. Like James Dunn and so many others, Schlesinger believed the medium might remake American politics and the individual's place in it. Your fireside chats, he wrote to the president, "marked an epoch in the history of democratic leadership." For William Nels, it took hearing the president speak over the radio to give him the sense that what took place in the halls of government mattered in his life. For Nels, and countless other listeners, radio enabled him to feel that he was a part of the national political arena. Writing Roosevelt only a few days after Nels, Schlesinger echoed the West Virginian. The president's radio addresses, Schlesinger said, "made me a part of government as I had never been before. It brought that cold abstraction, civic responsibility, down from the clouds and gave me an intense interest in public affairs. Public affairs became, in my mind, my own personal affairs."[55]

Something is lost in a democracy if we come to see public affairs in purely personal terms. The radio democracy that emerged from the 1930s was one often practiced apart from other people, in which democratic participation could mean private spectatorship. That should make us uneasy. But for Americans in the Depression decade, something had been gained. In their new ideas of democracy, listeners found ways to function in their modern mass politics and mass culture more generally—in a public sphere that seemed too large and too distant for individuals to influence. The mass communication of radio, of course, helped create this troubling world. But the mass culture that a public intellectual like William Orton worried would whittle away the individual could cut both ways. By personalizing the impersonal mass, by making an unfamiliar public familiar, individual listeners also used radio to help them carve out recognizable footing in their new world.

4

RADIO'S CHAMPIONS

Strange Gods?

Across the South and West a voice said, Come to me and I will heal your ills. And listeners came. Throughout the Northeast and Midwest a voice said, Cheer me and I will smite your enemies. And listeners cheered. From Washington, D.C., a voice said, Follow me and I will lead your cause. And listeners across the nation followed. The voices on the air could speak to listeners, many discovered. More than that, though, those voices could speak for listeners, many believed.

Up and down the radio dial in the Depression, men and women found champions on the air. In the voices they heard, many found speakers who stood up for listeners and their causes, heroes those listeners felt they could rely upon to fight for them. Radio offered some Americans the sense that they had a public say, a way to be heard in the vast national arena. Lone individuals in the twentieth century could rarely speak to the whole of their mass society. But on radio, a few special orators could reach a far-flung public. Because of the ethereal communities radio made possible, many listeners felt they shared personal relationships with those distant speakers. Those speakers were not simply distant and abstract voices, but spokespeople who felt like friends of sorts—friends with access to the nation's ears. Speakers such as Dr. John Brinkley in Kansas and later Texas, Father Charles Coughlin in Michigan and, most of all, president Franklin Roosevelt were, to some, champions standing in for, and up for, their supporters. To widely varying degrees, men and women—for good or ill—came to place faith in some of the voices on the air.

Here radio operated in paradoxical ways. The new medium was simultaneously empowering and manipulative. As radio fostered a feeling of personal connection between listeners and public personalities, many Americans used the medium to help them locate a voice in their mass culture. By finding powerful proxies to speak for them, some listeners gained a sense they might influence the increasingly large, centralized, and abstract public world. But as listeners latched on to their on-air advocates, they helped to create the very problems those listeners used radio, in part, to ameliorate. Because radio allowed only a select few places on the national airwaves, broadcasting helped further concentrate influence in the hands of distant speakers. This increased the risk that particular speakers might manipulate the public, eroding listeners' autonomy. At the least, the authority some radio speakers garnered further complicated the meanings of free speech, radio's democracy, and the place of the individual in modern America.

Listeners in the 1930s recognized some of these tensions. Millions felt they gained a public voice through the words of their on-air crusaders. At the same time, to those who disliked what a radio orator had to say, broadcasting appeared threatening. Listeners uncomfortable with a charismatic speaker's message correctly noted that radio also enabled those speakers to accrue power for themselves, often at the expense of their audiences. On the radio, the line between messiah and menace was, in fact, drawn in the air. Those intellectual critics of radio who suggested that the medium inhibited public speech—thinkers such as James Rorty and Ruth Brindze—were not entirely wrong about broadcasting and the meanings listeners found in it.

Whether we tend to see mass communication creating voices of democracy or demagoguery depends in part on just who the speakers were that listeners found so enthralling. President Franklin Roosevelt, Father Charles Coughlin, and Dr. John Brinkley were hardly the only champions listeners found on the airwaves, but they were some of the most widely known radio orators. Outspoken radio personalities dotted the dials as broadcasting developed. Like Roosevelt, Coughlin, and Brinkley, a range of speakers deliberately and explicitly offered themselves as defenders of their listeners' interests. Roosevelt, Coughlin, and Brinkley may serve as examples of several of the many types: politician, religious leader turned activist, and entrepreneurial doctor/medical savior—radio roles also played by the likes of Huey Long, Gerald L. K. Smith, Aimee Semple McPherson and Norman Baker among others.[1] Few speakers in such roles, though, were better known and more popular—or, in other circles, more reviled—than Roosevelt, Coughlin, and Brinkley: Coughlin and Brinkley attracted more listener letters to the federal commission overseeing radio than any other speakers. All three took advantage of mass culture's reach and grasped unusually widespread audiences. Radio made that possible, but it also constrained who could speak to that whole; access to the nation's ears was, of course, very limited. For Coughlin and, more dramatically, Brinkley, reaching that audience took considerable effort, but all three spoke to a mass public. Taken together then, the trio represents an extreme illustration of the many types of champions listeners found on the airwaves.

The stories of Roosevelt, Coughlin, and Brinkley reveal the meanings of such champions: the complex ways in which some listeners found in radio the means to be heard in a mass culture and how that process further centralized power, expanding that culture and at times muffling ordinary voices. To widely varying degrees, radio made all three heroes possible. All three used broadcasting to craft their popularity. To supporters, these figures took advantage of their access to a far-flung nation and far-off authorities to fight on behalf of their listeners. All three champions cultivated this image: Roosevelt put forth a very different vision than Coughlin and Brinkley, but each of them offered to help listeners in their struggles within the vast, impersonal, and interconnected society of modern America. The tremendous popularity such skilled speakers accrued provided them a dose of power as well. Supporters welcomed such influence: as long as their hero spoke with their tongues, fans believed that the voices on the air could amplify their own as well. But the balance was delicate. Many Americans saw Roosevelt, Coughlin, or Brinkley gaining influence as radio speakers and shuddered at the increasingly centralized and threatening control they gathered.

Indeed, even Roosevelt was among those shivering occasionally. The sway particular radio voices held over the public during the Depression made him uncomfortable—when he was not the one exercising it. And yet, radio's champions attracted large followings. Listeners sought something real. Americans wanted—and occasionally found in radio—ways to make themselves heard in the 1930s. Despite their opposing stands, those who criticized and those who applauded radio's champions often shared this common ground: both worried about the place of individual efficacy in their complex and abstract public arena. Clearly, the personal bonds listeners felt with these speakers and the ensuing sense that these speakers fought for their listeners were at least somewhat—and at times dangerously—illusory. But not entirely. Only when speakers could convincingly claim to speak for a significant section of the public did they win an audience.

As a mass society emerged in modern America, the question of how to count in a national public sphere was a real one. That those listeners who turned to radio champions found sometimes contradictory answers was, perhaps, not so surprising. To Roosevelt, the uncertainty of the era shaped listeners' search for radio heroes. "In normal times the radio and other appeals by them would not have been effective," he declared in 1935. "However, these are not normal times; people are jumpy and very ready to run after strange gods."[2]

Radio Stars

Certainly many, many listeners were ready to run after the voices they heard on the radio. Despite major differences in their presences, Roosevelt, Coughlin, and Brinkley all built their influential public lives, in part, through radio. Listeners came to know the president, the priest, and the physician through the air; and large numbers of those listeners came to support those radio voices devotedly. Like other radio orators, these speakers built huge bases of support with their broadcasts. Roosevelt was, of course, president, but his use of radio helped to make him beloved as well. Without radio, Coughlin and Brinkley might have spent their lifetimes in the bush leagues; with it, the parish priest and country doctor leapt into the majors and were known across the nation.

Serial dramas and adventure stories, comic variety programs, music, and the like were not the only genres of popular broadcasting in the late 1920s and the 1930s. Forceful speakers, delivering a wide variety of talks, speeches, and lectures could attract large audiences as well. Politicians obviously used radio to reach listeners nationwide at times; and although not always heard on network radio—and hence less often heard across the nation—an array of more controversial speakers regularly found their way into America's homes as well. Roosevelt's fireside chats commanded unmatched audiences. In the year before the United States' entry into World War II, well over half the homes in the country—between 40 and 50 million families—tuned to his chats. Coughlin's and Brinkley's audiences did not equal the president's, but unlike Roosevelt and other radio politicians who took to the microphone only occasionally, the priest and the doctor were regular on-air personalities. Week after week, listeners heard Coughlin and Brinkley. Also unlike Roosevelt, neither had access to the national networks of radio stations. Nonetheless both found ways to reach receiving sets in significant portions of the country: Coughlin through his own network of local stations in the Northeast and Midwest; Brinkley through an exceptionally far-reaching station that blanketed much of the Midwest and later the Southeast and Southwest as well. "Coughlin," *Fortune* magazine reported, "is just about the biggest thing that ever happened to radio." Estimates placed the priest's weekly audience at between 10 and 40 million listeners. As for Brinkley, even before he secured his very high-power transmitter, a 1930 *Radio Digest* poll found Brinkley's station far and away the single most popular in the country. When Roosevelt, Coughlin, and Brinkley talked, people listened.[3]

To Roosevelt, that mattered greatly. Radio enabled the president to take his agenda to the public. By doing so, the president won over many of his listeners. The intimate connection a politician using radio could forge with the members of a mass audience gave a single skilled politician remarkable sway over public opinion and political loyalties. And Roosevelt was a tremendously skilled radio politician. Through radio, then, Roosevelt could place his message before his audience and mold their views. Hearing his talks, many listeners bound themselves to the president.

Listeners found Roosevelt's radio presence strikingly persuasive. Many suggested that each time the president took to the air he convinced thousands, perhaps millions, of voters of his cause. Even when the president was under attack or speaking to unsympathetic listeners, many found themselves swayed by Roosevelt's voice on the air. The aerial assaults on the president by Coughlin and senator Huey Long in 1934 and 1935 caused Houston's R. Rohrbough to doubt Roosevelt's policies. In April 1935, though, Roosevelt delivered a fireside chat and won Rohrbough over completely. "The friendly tone of your familiar voice and your utterances brimming full of self confidence . . . has caused my faith in you to return to its peak," he wrote. Even some self-identifying conservative Republicans claimed that listening to a fireside chat convinced them to support Roosevelt. "That talk did as much as any one thing to make us forget party differences," wrote Republican H. L. Boyer of Geneva, New York, after Roosevelt's chat at the end of his first week in office.[4] Radio seemed to listeners a political tool without peer; they raved about Roosevelt's never-before-matched presence.

That presence enabled Roosevelt to build up an unmatched level of personal loyalty among many Americans. Millions of Americans looked with devotion upon Roosevelt as a paternal leader. Radio played an important role in fostering that sense of intimate allegiance to a public figure. The president understood radio's potential here, but perhaps even Roosevelt was at times surprised by the depth of the devotion his use of radio inspired. Like many others, W. D. Dixon of Springfield, Massachusetts, claimed to be so moved by a fireside chat that he offered to sacrifice his right arm for the president. In all probability Dixon intended the offer as merely a figure of speech, but the figures of speech themselves are revealing. Listeners expressed their eagerness to serve Roosevelt, often in such personal terms. Some, writing the president, referred to themselves and their neighbors as "your people"—not simply as political supporters or fellow Democrats. Roosevelt's radio talks so entranced Margaret Englemann of Portland that she looked upon the president with unabashed awe. "Something of his strength and nobility issued from the obedient instrument and thrilled the intent listeners, akin, I imagine to the English cottager's thrill when the Prince of Wales knocks on his door," she gushed. But if some among Roosevelt's listeners regarded themselves as his vassals, they did so because they, like so many of his listeners, loved the president. Through radio he garnered personal as well as political loyalty. "Mr. President," wrote Kate Carmichael from Goldsboro, North Carolina, "you have it within your power to bind the people of the nation to you personally in a loyalty beyond that accorded any other leader of the country since Washington. That warm human interest note that irradiates from your voice over the radio goes straight to the heart of the average man and woman."[5]

The cases of Coughlin and Brinkley tell similar stories. Like Roosevelt, Coughlin and Brinkley displayed tremendous on-air charisma and helped pioneer the use of radio. Like Roosevelt, Coughlin and Brinkley translated their radio presences into impressive popularity with devoted listeners. In the arcs their careers as radio heroes followed, though, the pair had more in common than the trio. Coughlin and Brinkley are never linked in the standard historical accounts of the 1930s. Coughlin is often painted as a combination of a demagogue with a racist and fascist agenda, a populist spokesman pushing American politics to the left, or an advocate for a passing world, trying to dam up the stream of changes washing over America. Brinkley is more often forgotten, or remembered only as either a radio pioneer or the most flamboyant of a dying breed of medicine show charlatans.[6] But the Catholic priest from the urban, industrial North and the entrepreneurial doctor from the rural Bible belt shared related cultural roles. Far more than even the radio president, both built their public selves out of the air. Both won widespread fame and influence only when they brought their diverse calls for rebellion and their cloaks of religious righteousness to radio. Over time, both turned to politics: positioning themselves as outsiders, both used their radio popularity to drum up consciously populist campaigns. And finally, in the late 1930s and as World War II approached, both Coughlin and Brinkley revealed the limits of their radio-made influence, as they lost the faith of many Americans. For much of the Depression, though, broadcasting served both speakers as the source of their mass popularity.[7]

Coughlin's rise to national prominence, observed news commentator Raymond Gram Swing, was the story of a man inventing himself through radio. Coughlin used radio to present himself to a mass audience, winning popularity in the process. The priest's cultural status cannot be separated from broadcasting; unlike Roosevelt, Coughlin the public figure was Coughlin the radio voice. When the radio priest first took to the air in 1926, he broadcast religious sermons on a local station near Detroit. By 1930, however, he had achieved such regional popularity that CBS picked up his talks on its national network. With the onset of the Depression, Coughlin turned his attention to political and social issues, challenging communism, the international economic system, and president Herbert Hoover. CBS quickly found Coughlin's addresses too inflammatory for its tastes and in 1931 refused to renew the priest's contract. But Coughlin's national standing had already been secured. Although NBC also found the priest too controversial to air, Coughlin had little difficulty locating individual stations across the Northeast and Midwest that were eager for his voice. In an innovative—and expensive—move, Coughlin bought time from those stations and rented telephone lines to connect them during his broadcasts. It cost roughly fifty thousand dollars a week in the Depression, but essentially he set up his own temporary network of stations. By the middle of 1932 he had over twenty stations on line; by 1935, over thirty.[8]

On his makeshift network, Coughlin garnered a tremendous following. He attracted larger estimated audiences than any other regularly aired program, and those listeners responded to Coughlin in record numbers as well. The priest's voice did not reach west of the Rockies until after 1935 or into the South at all, but listeners flocked to Coughlin's broadcasts in many of the nation's urban areas. In the small city of Brockton, Massachusetts, referees interrupted high school football games if Coughlin was on the air so fans and players could listen. During the first half of the 1930s he received roughly eighty thousand letters each week—more than any other person in the United States. He employed nearly a hundred clerks to handle his mail. Listeners backed up their devotion to Coughlin by funding his radio enterprise through voluntary contributions. Millions of listeners demonstrated their loyalty to the radio priest by sending him small contributions on a regular basis; at his height, Coughlin received about five million dollars a year from listeners. Much of that, of course, went to pay his enormous expenses, but Coughlin still managed to parlay his radio celebrity into a near-million-dollar fortune during the Depression.[9]

For much of the 1930s, Brinkley, like Coughlin and Roosevelt, held a significant audience in thrall. Like Father Coughlin, Dr. Brinkley created himself for a mass public almost entirely through radio. Part evangelical doctor, part flamboyant businessman, part con artist, Brinkley used his radio station—and some highly dubious medical practices—to turn himself into a millionaire, a serious political candidate, and a hero for many in rural America, especially in the Midwest and South. Despite several organized efforts to discredit and silence the doctor, Brinkley gained a large popular following. "Whether he is a fake or not, I have no way of knowing," wrote a skeptical

Ernie Pyle, the traveling journalist. "But if he is, I wish he would give me some lessons in fooling the public."[10]

Without radio, Brinkley likely would have been little more than one of the many medicine-show doctors done in by the twentieth-century movement to professionalization. Born in rural North Carolina in 1885, Brinkley bridged two eras. He had seen his father practice medicine as a country doctor, dealing directly with patients, largely unregulated by governmental and professional boards. From the start of his own medical ambitions, though, such organizations shaped Brinkley's career. Without money or the academic background for a well-regarded medical school, Brinkley took his training either at questionable schools or in apprenticeships. After settling in tiny Milford, Kansas, in 1917, Brinkley developed the operation that earned him the title "the goat-gland doctor." As Brinkley told it, a farmer came to the doctor complaining of impotency. Looking at several goats mounting each other, Brinkley joked that the patient would have no trouble with sex with those goats' sex glands in him. The patient insisted that Brinkley perform the transplant. He did, and within a year the farmer's wife gave birth to a boy, Billy. For years afterward, Brinkley offered his goat-gland transplant for nearly any problem that could be classified as "failing manhood."[11] In the mid-1930s Brinkley abandoned transplantation for other, equally questionable procedures, but the goat-gland reputation stuck with him.

Brinkley's reputation not only stuck, but spread far beyond Milford and Kansas. It did so because he publicized himself through the radio. In 1923 Brinkley founded the station KFKB and transformed his career. Although his station aired a wide range of material, over the course of the 1920s Brinkley used it more and more to promote his medical practice and himself. In addition to broadcasting music, religious discussions, and the *Tell-Me-a-Story Lady* program, six days a week Brinkley delivered his own lectures on the air. Speaking intimately and colloquially, Brinkley sought to convince listeners to take their health in their own hands and come to see him. In 1929 he started what would become his best known program, *The Medical Question Box*. Listeners wrote to Brinkley describing their medical problems and Brinkley analyzed the ill on the air, offering prescriptions and other recommendations. By the start of the Depression he had contracts with pharmacies throughout the region to sell his medicines. He set up a numbering system so that when prescribing for a patient by radio, Brinkley simply told listeners to buy his medicines by number. It minimized confusion, he said. It also ensured that listeners bought his highly questionable brand of drugs and helped to make Brinkley a familiar presence.[12]

Brinkley's radio talks and on-air medical advice may or may not have done much for the health of his patients. But, as it did for Roosevelt and

Coughlin, broadcasting brought him a large and dedicated following. At least one family of Brinkleyites reserved the space above their mantle for the doctor's picture; and more than a decade after the doctor's death, the name Brinkley was still revered in homes across the plains, a researcher found. Although he never broadcast on network radio, KFKB's signal was powerful enough to cover much of the Midwest. Listeners regularly tuned in to hear the doctor from as far away as Chicago. The federal government tried to silence Brinkley's station, but that only broadened his audience. By 1932 he had set up a new, exceedingly powerful station in Mexico. That station, XER, had the strength to reach listeners as distant as New England or California, depending on weather conditions. Many listeners found Brinkley's talk about health and sex offensive, but many others tuned in regularly. *The Medical Question Box* drew as many as thirty-five thousand letters from listeners in a week. Brinkley employed a staff of fifty just to handle his mail. Brinkley clubs sprung up across the region; by 1931 the children's spin-off of the Brinkley clubs, Cowboy and Cowgirl Ranch Clubs, numbered 224. And listeners flocked to his hospital too: in the five-year period from 1934 to 1939 Brinkley treated fifteen thousand patients, an estimated 90 percent of whom were attracted by his radio talks. C. P. Collins of Bellefontaine, Ohio, believed that Brinkley's radio broadcasts had saved his life. And he believed in Brinkley. "I am 69 years old now in first class health all caused by having a Radio, and i tuned in when this Doctor was broadcasting," Collins wrote to president Franklin Roosevelt. "If this Doctor had not had the privelage [of operating], no doubt the undertaker would have closed the case long ago."[13]

Voices of the People

C. P. Collins supported Brinkley, at least in part, because he believed the doctor had physically done something for him. Most listeners, of course, never actually met radio speakers such as Roosevelt, Coughlin, or Brinkley. Yet pieces of Collins's sentiments would have rung true with many of the trio's supporters. Listeners tuned into these orators by the millions, sent them letters by the tens of thousands a week, and invested themselves—often emotionally or monetarily, occasionally even physically—in their words. Some may have listened largely for entertainment or information or even to be outraged. But the devoted popularity of these speakers depended on something more. Listeners latched onto a Roosevelt, a Coughlin, a Brinkley, in part because such speakers offered listeners something they desperately wanted in the 1930s. Collins believed Brinkley had worked on his behalf in the operating room. A good swath of listeners believed such a speaker worked on their behalf in modern America more generally. To supporters, the president, the priest, and the physician spoke forcefully for the people. They gave listeners a sense of having a voice in a centrally controlled and vast mass culture.

Radio, of course, made possible, and indeed abundant, ethereal relationships, ones in which listeners forged personal ties with voices they felt they could count upon to help them when their face-to-face resources were inadequate. Janet Bonthuis, remember, sought a Christmas present for her father from *Vox Pop* announcer Parks Johnson when she lacked the money to buy one.[14] That some listeners would fashion and count upon similar ethereal relationships with the likes of Roosevelt, Coughlin, and Brinkley among others is hardly surprising. What made relationships with such charismatic speakers different, though, was that these figures deliberately presented themselves as speaking for and, indeed, crusading for their listeners. As Roosevelt, Coughlin, and Brinkley often overtly took on the pose of champion, some listeners came to see all three representing the common people amidst the distant systems that characterized modern America.

Perhaps most evidently, to Roosevelt's supporters, the president embodied the public. The actions he took were on behalf of those listeners. Roosevelt's overwhelming success at the polls proved that the president spoke for the people, explained Bernard Sachs of New York City. Consequently, Sachs, continued, Congress should not oppose the president: to do so was to oppose the people themselves. Those listening to Roosevelt believed that his strength empowered citizens who felt no means of affecting national politics. The radio president, they believed, could bring the citizenry into the halls of power. Common citizens alone had no voice on Capitol Hill, lamented W. A. Endy of Chicago. But Roosevelt, he said, could speak for them anywhere and, with radio, everywhere. And the more authority Roosevelt accrued, the better he could represent listeners, his supporters felt. Consequently, the more the ability to speak on the air empowered Roosevelt, the more the will of the people would reign and the better democracy would function. "You are the chosen captain," Benjamin Weis of Columbus, Ohio, declared "and Congress is expected to enact legislation in accordance with your demands."[15]

Popular rule in modern America, suggested some, could best be achieved by concentrating power in the hands of a leader who would crusade for the populace. In Roosevelt, many believed they had found such a champion of the people. Or, as writer John Dos Passos described it, those who heard the president's fireside chats got the sense that Roosevelt was "operating the intricate machinery of the departments, drafting codes and regulations and bills for the benefit of youandme, worried about things . . . for the benefit of us wageearners, us homeowners, us farmers, us mechanics, us miners, us mortgagees. . . . The machinations of chisellers are to be foiled for youandme

. . . the manycylindered motor of recovery is being primed with billions for youandme."[16]

This was a deliberate effect on Roosevelt's part. The radio president repeatedly and overtly presented himself as the embodiment of the people. "The people of the United States," Roosevelt declared at his first inaugural, "have made me the present instrument of their wishes." In Roosevelt's eyes, he, more than anyone else in the country, understood both the problems facing America and what the people wanted. He repeatedly reminded listeners that he shared their struggles. "It is your problem, my friends," said Roosevelt, concluding his very first fireside chat, "your problem no less than it is mine." Obviously Roosevelt used radio as an extensive propaganda machine, but he did so by speaking directly to listeners' real concerns about their lack of efficacy. And he offered his radio persona as a mediating force. Starting with that first fireside chat—when he explained the American banking system and how the government was reforming it in plain, human terms—Roosevelt suggested that he personally could help citizens understand and, if indirectly, participate in the actions of their government.[17] As we saw in the previous chapter, Roosevelt helped revitalize the idea of popular government by embodying a new view of democracy, one adapted to the massive bureaucratic state. He held himself up as a conduit through which listeners could personally touch an otherwise distant and indifferent political arena.

Coughlin—despite his claims—never attracted the same level of widespread devotion as Roosevelt, not even in the first half of the 1930s when the radio priest's popularity peaked. But his followers similarly looked to Coughlin to speak for the people. Thanks to the priest's affinity with his listeners and his access to the radio waves, his supporters believed, their interests gained a meaningful public hearing. To Coughlin's admirers, he stood squarely with them. "It is," claimed Mary O'Gorman of Toledo, Ohio, "an undeniable fact that the rank and file of the American public is heart and soul with Father Coughlin in all his utterances." Like many who wrote to the Federal Communications Commission and Coughlin's critics, O'Gorman felt a personal dependence on the priest. She could rely on Coughlin because not only was the priest in touch with her rank and file, but because she believed that radio enabled him to speak for them. Dorothy Parker of Highland Park, Michigan, dreaded the possibility that federal radio authorities might close the airwaves to the people's champion. "I beg that Father Coughlin be not taken from us," she pleaded. "He belongs to us." Coughlin, she explained, was an attorney, standing up for and defending his clients in a world with regulations they could not understand. "He is our spokesman and protector. We need him and feel that we have a right to him. He speaks to us and for us," she wrote. "Why should the multitudes be deprived of their protector. . . . This country belongs to the multitudes. Father Coughlin belongs to the citizens of these United States."[18]

Coughlin presented himself in just such terms. He identified himself to listeners as "your spokesman." His authority, he declared, came from his unmatched ability to express popular desires. "I am not boasting when I say I know the pulse of the people," Coughlin crowed. No one received more mail than he did, Coughlin explained to prove his point: the thousands of letters he received daily taught him just what listeners felt, he said. Those letters, he argued, were not only a testimony to his popularity, but gave him his special ability to represent the public. He considered his millions of listener letters to be "the greatest human document of our time"—the greatest document and he possessed it.[19] Looking back on Coughlin years later, we cannot miss his racist and fascist sympathies, but we must also remember that he built his influence through his claim to be a conduit of the public will, the people's champion. And the new mass medium of the era helped make that claim convincing to millions of Americans.

Millions of American believed Coughlin stood with them, outside the mainstreams of authority that interlaced their world. The priest not only gave his listeners a voice, as he did so, he articulated their sense of disconnect from those in control of modern America. Mary O'Gorman, who believed Coughlin walked hand in hand with "the rank and file," revered the priest for battling against those in power on behalf or ordinary Americans. Similarly, when Marion Orr of Hamden, Connecticut, cheered Coughlin for fighting for the common people and American values, she praised the radio priest for attacking an insidious conspiracy mounted by abstract forces of big business.[20] The listeners who cheered Roosevelt frequently did so for humanizing the impersonal centralized systems of power that daunted them; those who lionized Coughlin often applauded him for trying to tear them down.

Coughlin spoke to the worries of many of his listeners by addressing several decades of dislocating changes. He expressed their frustration, lashing out at the centralization of power and a world of large, bureaucratic organizations that left little room for traditional or meaningful individual autonomy. "Hidden forces," the priest claimed, "have conspired against the common people of the world." Through more than a decade of broadcasting, Coughlin did not always maintain a clearly coherent ideology. He was capable of blasting massive institutions or attacking the Tennessee Valley Authority as socialism at one moment, and then calling for the nationalization of valuable resources and the expansion of the government to referee his Jeffersonian utopia the next. But whether he attacked the international banking system, large-scale capitalism, or communism in general, Coughlin gave voice to many Americans' concerns about being disempowered in the mass society emerging in the early twentieth century.[21] To his supporters, he seemed to offer a means of restoring face-to-face values to the modern world.

The deep devotion another, probably somewhat smaller, set of listeners felt for Brinkley stemmed from similar sources. Many of those who made Brinkley popular followed the doctor because he seemed to them to represent ordinary Americans, and to represent them in opposition to distant controls operating through impersonal bureaucracies. Supporters claimed that Brinkley seemed to know them personally and speak for them individually. Each day, many said, he entered their homes through the radio and exactly described their symptoms and woes, medical or otherwise. In 1930 the federal government began a decade-long campaign to close the airwaves to Brinkley—and in the process, inadvertently helped shape some listeners' close doctor-audience relationships. The American Medical Association and the Federal Radio Commission believed the doctor was defrauding the public and sought to put a halt to his medical and broadcasting activities. The AMA attacked Brinkley on the legitimate grounds that he had not been trained at an accredited institution and that his goat-gland operation lacked scientific validity. In September 1930, the professional organization revoked his license to practice medicine in Kansas. With the AMA urging it on, the FRC closed the airwaves to KFKB late in that year. Prescribing medicine over the air and medical talks by an unlicensed doctor, the commission ruled, did not serve the public interest—the legal standard for a broadcasting license. To his fans, though, this helped confirm their sense that Brinkley embodied the will of ordinary folks. When Ethel Rauber of Newton, Kansas, later wrote to Roosevelt in a plea to keep Brinkley on the air, she suggested that popular rule could be found in the voice of the radio doctor, not in national politics. "Politicians," Rauber told the president, "strangle the people even under your New Deal."[22]

That Rauber equated an attack on Brinkley to an attack on "the people" was, of course, no accident. As Brinkley cast himself, he sprang from the common people and found his strength and direction in them. The biography he authorized about himself took the assumingly unassuming, everyman title of *The Life of a Man*. Brinkley made it clear he had risen above his poor rural background by ostentatiously displaying his success, but he also professed to channel the will of the masses. His radio station, he claimed, was the people's station—one locally controlled, rather than a network affiliate serving as a vehicle for a distant cultural elite. The call letters KFKB—often translated into Kansas's First, Kansas's Best—stood for Kansas Folks Know Best, Brinkley asserted. In 1930 he ran for governor and promised to put a radio microphone in the governor's office. The voters, he said, deserved the full confidences of their political leaders. The doctor was, his biographer Clement

Wood wrote, simply the people's voice. From Wood's perspective, though, no higher calling existed: "John R. Brinkley is not important in himself: no man is. The most that the highest among us can achieve is to be the topmost drop on a huge human wave rolling in some unnoticed direction."[23]

To his supporters, of course, Brinkley was more than simply one drop in the ocean of humanity. In his battle with the government of the United States, Brinkley seemed to demonstrate that the interests of those ordinary people should and could triumph against centralized modern authority. In Brinkley's efforts to stay on the air, his listeners saw him manifesting their own frustrations with faraway rules and systems of power—and finding a way to win. When the Federal Radio Commission pulled his broadcasting license in Kansas, Brinkley relocated to Mexico, beyond the central regulator's reach. To his supporters, the doctor's powerful Mexican station, XER, represented an alternative to network broadcasting, to control of American culture by New-York-centered big businesses, and a slap in the face of the federal government's efforts to dictate airwaves all over the country from Washington, D.C. "It is a disgrace that the air should be monopolized By the big interests like N.B.C. + C.B.S.," wrote Fred Reilly of Burlingame, California, explaining to the Federal Radio Commission why he cheered Brinkley and his station. Reilly made the effort to write to the government supporting Brinkley because he believed in the doctor's independent cause; Reilly had no special fondness for the particular programming on Brinkley's station, he claimed. The doctor's supporters seemed to feel they shared in his success, that his continued radio presence represented a personal victory of sorts. E. A. Boehme, a farmer from Offerle, Kansas, found Brinkley's broadcasts from Mexico a sign that Boehme's values and the individual more generally could best impersonal central controls—of all kinds. "Well what do you think of X.E.R. We shure do think it is fine. The most popular station in the *whole world*. So he can't talk over Radio ha! ha! Yours for less chain," Boehme crowed to federal radio regulators. "Down with the Farm Board and Radio Com."[24]

This seemingly odd juxtaposition of "the Farm Board and Radio Com" is a revealing one. When Boehme cheered Dr. Brinkley and berated the Farm Board in nearly the same pen strokes, he hinted at the issues of the day he found most troubling. Boehme disdained the distant and impersonal authorities he felt intruding into his life. Brinkley, unlike Roosevelt and Coughlin, did not generally focus on overtly political material in his broadcasts; the radio doctor was, after all, primarily trying to sell listeners on his medical treatments. But he spoke directly to his followers' modern discomforts. As Coughlin did, Brinkley attacked the enormous institutions of twentieth-century America. Massive organizations, the doctor declared, empowered select elites at the expense of ordinary individuals. The many efforts to silence him, Brinkley declared, represented the work of large bureaucracies controlled by newly centralized authorities attempting to support their own interests at the expense of the common people. Brinkley's authorized biography declared that the American Medical Association threatened traditional American ideals of individual liberty. The AMA, Brinkley claimed, was jealous of his success and afraid for its status. According to his biographer, in fighting the AMA, Brinkley was "fighting for the people, against useless medication, against useless operations." Brinkley similarly assailed federal radio regulators and the political parties, accusing the political system of disregarding the welfare of the whole in its effort to benefit a few. On the air, he regularly lashed out at Wall Street's power to control the economy, even on Main Street. And he bemoaned the system of national radio networks, which, he claimed, centralized programming in New York City and ignored local tastes around the country. Not surprisingly, Brinkley, like Louisiana governor and senator Huey Long, was most popular in counties where Populism had reigned more than thirty years earlier. Like Long, who mined many of the same radio techniques as the doctor, Brinkley tapped into a resurgent vein of frustration with systemic economic hardship and bureaucratic organizations.[25] Repeatedly he told his listeners he stood with them against the growing octopus.

In his broadcasts, Brinkley offered listeners a way to fight back against the tendrils of impersonal and far-off power. He was, the doctor all but declared, a crusader fighting to empower his listeners. As he discussed health, he held himself up as a personal alternative to an impersonal and emasculating system. His medical services, he suggested, gave listeners a means of taking the control of their health away from professionalized doctors. By taking charge of their own prostates, Brinkley told his male listeners, they could reclaim their manhood. "Are you a man of your own mind? Or are you a weakling?" asked Brinkley. "Must you run to your doctor before coming to me? Who's prostate is it—yours or your doctor's?" Obviously, turning to Brinkley gave a patient no more self-control than relying on a local doctor did, but Brinkley did his best to convince listeners otherwise. On the air he often punctuated his talks with heavy sighs, as though he were weighted with the responsibility of the health of all listeners. "He is convinced that he was put on earth by God to fulfill a mission, that of healing the sick," Brinkley's biographer gushed.[26] Although the maligned local doctor might in fact live in the listener's vicinity, Brinkley suggested that such doctors were simply extensions of the faceless AMA; he, on the other hand, listeners knew personally from the radio.

Brinkley frequently alluded to himself as a metaphorical modern Jesus.[27] The comparison made sense to the radio doctor. As he struggled with the federal government and the medical profession, he cast himself as a martyr to the new centralized authority of modern America. And as he fought back successfully, he cast himself as a hero for the common people. To his supporters, such a stance felt authentic.

Roosevelt, Coughlin, and Brinkley did not all offer their listeners the same visions for America. Roosevelt, among other things, proposed to humanize and moderate some of the massive networks of power that tangled up so many Americans by the Depression, while Coughlin and Brinkley attacked the modern world more directly and claimed to offer alternatives to restore the potency of the individual. But to those who supported each of the three speakers, the president, the priest, and the doctor gave listeners something they desperately wanted: a means to get their views heard in a mass society. To their supporters, their stances as the people's champions felt authentic.

Power . . . Corrupts?

To those listeners who did not like particular speakers or their messages, though, such speakers only posed as heroes of the people. And even at the heights of their popularity in the 1930s, many disdained or despised Brinkley, Coughlin, and even Roosevelt. To their opponents, radio looked very, very dangerous. Charismatic speakers used broadcasting to secure tremendous popular support; and that support, in the right hands, translated into power. But for whom? Certainly radio empowered those few talented speakers who could produce programs at a single location and broadly distribute their words. Followers of the president, the priest, or the doctor welcomed that: as long as radio empowered their heroes—and their heroes represented the will of the people—listeners believed they could have their say. Opponents of radio's various champions, though, saw a problem here: what might look like representing the people, could in fact be a case of manipulating the masses. There is clearly an ominous side to the idea of a select few speakers gaining unusual influence in society. To the many Americans who did not like those particular speakers, radio appeared to give them a means to dupe and manipulate listeners from afar. This fear grew from the same soil as the concerns that spurred defenders of radio speakers: a fear of individuals losing control of their lives. Broadcasting helped foster the very centralized culture the likes of Coughlin and Brinkley won popularity fighting against.

Through radio, Roosevelt, Coughlin, Brinkley, and similar speakers indisputably transformed mass popularity into a source of personal power. This was something new in the twentieth century. On the air a skilled speaker could reach and, importantly, sway thousands or even millions of listeners. Moreover, by threatening to mobilize those listeners as a group, successful radio champions could make sure their calls echoed through corridors of power far from those listeners. Popularity need not mean fame alone; with the instantaneous and personal connection broadcasting allowed, celebrity could readily translate into clout.

It was in no small part through his use of broadcasting that Roosevelt empowered himself and the office of the presidency. Listeners in the 1930s clearly understood that through radio, Roosevelt had changed the balance of power in Washington, using his influence over public opinion to place himself and the executive squarely at the center of American politics. Radio, said many in the 1930s, would be Franklin Roosevelt's big stick: as long as he did so through the right medium, the president could speak softly and amass the people's support to prod Congress or big business into line. "As long as you talk to your people there is not one thing you cannot accomplish," wrote F. B. Graham of Dubuque, Iowa. "Congress and other law-makers will find themselves puny interference when you have but to turn to the Radio and enter our home a welcome and revered guest." The microphone, listeners asserted, was Roosevelt's greatest weapon. Keep in mind that polling designed to determine national moods only began to develop in this 1930s. Prior to radio, as many listeners in the 1930s understood national politics, public opinion was murky and had little impact. But as Roosevelt used the medium, many felt, he brought public opinion into the political arena and made it into a directed force. This felt entirely new. To listeners, it opened up a fresh realm of authority for the president. Through radio, Carl Geigand of Buffalo explained, "influences, adverse to your wise legislative policies, can be brought into proper alignment since the power of public opinion will be thrown in the balance."[28]

Precisely because many listeners applauded politicians' speaking on radio, this shifted the nature of political influence. As radio took hold, power increasingly stemmed from popularity and lay with the president. Although it would become commonplace in broadcasting's era, the idea that a president might appeal to public opinion—might play to his popularity—as a source of authority was largely a twentieth-century innovation.[29] Listeners in the 1930s, newly familiar with their president through radio, repeatedly begged Roosevelt to take to the air more often. He could, they believed, give the people a voice by organizing them to apply pressure on those who opposed his agenda. By using radio to rally the country, listeners claimed, their political champion would cower his opposition and accomplish their shared agenda. Clearly the personal connections and appeals radio made possible gave the president special status. Radio favored a leader who could reach the whole nation; and the president was the one elected politician who could claim to be responsible to the entire country and who had unmatched access to the networks' airtime. As Roosevelt took advantage of that, he expanded the authority of his office. A listener like Geigand admitted he looked to the executive for *legislative* policies. And after hearing Roosevelt speak, a listener like Graham was more than willing to see the president sweep away the powers of Congress.

Radio speakers did not need to hold elected office to put their popularity to use. Coughlin never had the political influence he claimed, but his connection to and sway with the public also made him a formidable political force. In 1935, for example, with the Senate likely to ratify the treaty committing the United States to join the World Court, Coughlin took advantage of a weekend recess before the vote to blast the plan on his Sunday radio address. Listeners, following Coughlin's cry, jammed the telegraph lines into Washington, D.C., with protests against the treaty. Where only a few years earlier a dozen telegrams on the same subject would have impressed Congress, Coughlin's radio address produced tens of thousands of protests. On Tuesday, the Senate rejected the treaty. Radio had enabled Coughlin to relate to the mass of the public as no popular leader before the 1930s. The medium was the source of his popular appeal, and, as his own statements made clear, that appeal was the source of his clout. Coughlin appreciated political muscle and through radio he toned his.[30]

Brinkley also successfully used radio to bolster his own authority, but he sought influence of a different sort. Unlike Roosevelt and Coughlin, the doctor did not focus on national politics, but pursued more personal and local interests. In doing so, though, he clashed with national officials and used his broadcasting power to fight back. He converted his large audience into a fortune and, very nearly, into political office. The doctor persuaded listeners to invest millions of dollars in his words. During the Depression Brinkley's income topped a million dollars in a year. For Brinkley, wealth was a key measure of his success. He turned his radio popularity into fancy cars, several yachts, a plane, and an ever-worn array of diamond jewelry. The diamonds, no doubt, were supposed to impress, but it was not the rocks alone that led the Milford pastor to see Brinkley with awe. He was, the pastor said, "a man of destiny, a great man—the greatest I ever met."[31]

Brinkley's radio persona did not, however, win over everyone. When the American Medical Association and Federal Radio Commission worked to close his Kansas medical practice and radio station in 1930, Brinkley responded by exercising his influence on a larger stage. With his public persona and livelihood in danger, Brinkley fought back, turning his celebrity into political power. In a testimony, at least in part, to the widespread popularity of his radio image and his ability to draw upon that status, Brinkley very nearly grabbed the governorship of Kansas in a whirlwind campaign. Under fire from the AMA, in September 1930 Brinkley announced his candidacy for governor in the November election. Too late to get his name on the ballot, Brinkley ran as an independent write-in candidate. Blitzing the state with radio addresses and flamboyant camp-meeting-style rallies in which he was as likely to talk religion as politics, he spoke to the frustration of the early Depression. As he did on his regular radio broadcasts, he presented himself as a common representative, fighting back against the oppressive established system: the AMA, and the Republican and Democratic political machines. Only some clever ballot counting thwarted the radio doctor's evangelical campaign. The state body overseeing the election decreed that only write-in votes for J. R. Brinkley would count toward the doctor's total: ballots marked John Brinkley, Dr. Brinkley, or the like would be discarded. Despite a concerted effort by Brinkley and his clubs around the state to inform voters about the regulation, as many as fifty thousand voters consequently had their ballots rejected. The write-in candidate lost to Democrat Harry Woodring 217,171 to 183,278. Woodring edged out Republican Frank Haucke by fewer than 300 votes, but Haucke did not call for a recount; Republicans feared a recount would show Brinkley as the actual winner. Although defeated, Brinkley demonstrated his popularity and his followers' enduring faith in him. Despite running against both parties and dubious election practices, he polled hundreds of thousands of votes—not including the tens of thousands of write-in votes he received in states other than Kansas.[32] Moreover, the Brinkley clubs gained members after 1930.

Electoral defeat and the AMA and Federal Radio Commission attacks did not do away with Brinkley. Indeed, as the decade continued, Brinkley used radio to expand his economic power, if not necessarily his political influence. Banned from the air in the United States and from practicing medicine in Kansas, Brinkley moved his operation to Del Rio, Texas, and set up a radio station just across the border in Mexico. Broadcasting from beyond the jurisdiction of the United States government, Brinkley had little difficulty reaching and extending his audience. Unfettered by American regulations, in 1932 his Mexican station, XER, was the highest-powered station in the world, transmitting with 500,000 watts. At a time when the average NBC station broadcast with 10,000 watts and the average nonnetwork station at 566 watts, and when only one station in the United States topped 50,000 watts, Brinkley blew stations off the air all over the United States. In Colorado, for instance, listeners said that they could no longer hear the Denver NBC station, KOA, because Brinkley broadcast on the same wavelength; the federal government heard similar reports from California to Pennsylvania. Despite the federal government's repeated efforts to convince Mexico to halt Brinkley's broadcasts, the radio doctor remained on the air. And he continued to use radio to boost his following, his booming medical practice, and, for a few years anyway, his political aspirations. In 1932, despite having effectively relocated to Texas, he again ran a close third for the Kansas governorship, polling 244,607 votes to winner Alf Landon's 278,581. As he became more involved in life in Del Rio, Texas, Brinkley withdrew from electoral politics, but that did not negate his influence. Broadcasting with superhigh power from Mexico brought him more listeners—and more followers willing to invest their money along with their faith in his hands.[33]

A big part of the reason Brinkley had to be so resourceful to get access to the air, of course, was that the Roosevelt administration, along with much of the public, sought to silence the doctor. The power that speakers like the president, Coughlin, and the doctor amassed through the air made some Americans very uncomfortable. By concentrating influence with charismatic speakers, broadcasting opened possible new avenues of manipulation. These paths might lead not to democratic rule, but to dictatorial authority. Radio, some feared, could give otherwise-isolated individuals the ability to foment national upheaval. Speakers on the air could well swindle, not speak for, their listeners. There was real tension here: in the guise of giving the people a voice, radio might also disempower individual listeners across the country.

Despite—or rather, in part because of—the impressive endorsements Roosevelt won at the polls and received in letters from listeners, the president's mastery of radio caused some to fear for the future of democracy. It created new—and, to some, dangerous—kinds of authority. Through radio, Roosevelt foes occasionally charged, the president amassed personal power and stifled opposition, potentially laying the groundwork for the sort of dictatorial rule rising in Nazi Germany. Roosevelt repeatedly used radio to cast a spell over America, hypnotizing listeners to secure their support, the anti-Roosevelt *Chicago Tribune* charged. With that extensive personal following, critics feared, Roosevelt subverted the proper balance of power in the United States: he could use his sway with the public to dictate legislation. A 1935 political cartoon showed Roosevelt coming through a radio to whisper in the ear of a woman representing public opinion. Enraptured by the radio speaker, she brandished a rolling pin while her husband, representing Congress, looked on in shocked fear. "Delivering a Message to Congress," the caption read. To a vehement Roosevelt critic such as Herbert Hoover, this had clear and forbidding implications. He did not see Roosevelt giving voice to the will of the people; he feared that Roosevelt managed the will of the people to win undemocratic powers for himself. "You cannot have government by public opinion when opinion of the people is manufactured by paid press agents of the government," Hoover told Kansas City Republicans in 1938. The molding of mass opinion—prejudice, he called it—spawned dictatorships. "It is the sustenance they live by," Hoover announced. "It is the stuff the New Deal lives by."[34]

Such concerns were not simply the stuff of political rivalry. Roosevelt himself recognized and at times worried about radio's manipulative potential. He took pains to avoid the sorts of connections Hoover found between the radio president and those with autocratic aspirations, in the United States and abroad. Roosevelt did not see himself misusing the public trust: his use of radio differed from that of Coughlin and Huey Long, the president asserted; and his presence was certainly different from that of Adolf Hitler, who used radio successfully in Germany. The president was, however, very aware that radio speakers could abuse the confidence of their audiences. And, like Hoover and an array of listeners, that possibility disturbed him. His administration feared that Coughlin dangerously misused his broadcasting power. Additionally, for much of the 1930s, the White House worked to shut down Brinkley's Mexican radio station. Indeed, in 1933 Roosevelt's ambassador to Mexico declared that silencing Brinkley was the most pressing matter for relations between the United States and its neighbor.[35] And his administration had plenty of company in worrying about the sway radio afforded those speakers over their large audiences.

Listeners who opposed Coughlin in his heyday in the early and mid-1930s typically feared that his radio diatribes would divide the United States and create internal strife. Broadcasting, they believed, was a dangerous tool; it gave a single individual too much power to stir up resentment. "One man today can do more harm over the 'mike' than a thousand will that do not use it," wrote Ernest Stillman, a New York listener who felt Coughlin had caused the 1933 banking crisis.[36] Like Coughlin's supporters, his opponents saw social and economic uncertainty presenting certain dangers. But unlike his supporters, those opponents saw radio making such matters worse.

Coughlin's opponents heard the radio priest using the mass medium to spur ethnic, religious, and class divisiveness and resentment. To one set of critics, radio enabled a speaker to play on petty divisions and spread them through the nation. Many listeners who disliked Coughlin argued he used his microphone to spread religious hatred, particularly, they, felt, either anti-Protestant or anti-Jewish feelings. "The spirit of unkindness, rancor and bigotry which characterizes these speeches is out of harmony with the generally constructive nature of American radio programs in general," wrote D. L. Browning of London, Ohio, one of many Protestant ministers to voice reservations about the radio priest. "The fomenting of religious strife is not only a useless but a dangerous thing."[37] Others made a similar argument about Coughlin's use of class and ethnicity; many believed, even as early as 1935, that Coughlin supported Nazi ideology. The priest, they worried, used radio to undermine the pluralistic tolerance needed for a diverse nation.

Another group of listeners critical of Coughlin were more concerned with maintaining their status and feared that the priest's use of radio would rip apart the United States by undercutting traditional hierarchies of class and religion. To these listeners, the risk was that, by using radio, Coughlin might bring together the working class or Catholics from across the nation and spur those groups to action. Some in traditional, white, Protestant America feared working-class unity and activism; and they saw the new mass medium in the hands of a Coughlin as a means to create such a new bloc and produce that effect. "He is reaching a class element that mite cause us a great concern," Baltimore listener William Wood warned. For many, resisting Coughlin was important because they saw him empowering Catholics, who, they believed, would inevitably degrade America's traditions. "Is it possible that the National Commission is afraid to cut him off because the speaker is a Catholic priest who happens to have a large following among the ignorant class of people?" asked R. B. Wood of Adams, Wisconsin. "Is he to go on until he actually foments riots and revolution?" Listeners like these were uneasy with social dislocations of the past several decades and they dreaded the possibility that radio, in the hands of a Coughlin, could increase such turmoil. Because, as George Clarke of Waukegan, Illinois, pointed out, radio "embraces all classes of people," it could be used to rouse members of the working class across the country and point them to action.[38]

This group of critics often saw Coughlin creating dissatisfaction in particular groups rather than giving voice to their existing impulses. George Clarke went on to suggest that "each and every one of these people who listen are consciously and unconsciously affected by what is said or implied by speakers on the air." Listeners, in other words, could not help but be at least somewhat influenced by what they heard.[39] Coughlin, this group of opponents feared, might use his radio presence to prod otherwise inactive groups to action. Prejudice clouded the vision of some of Coughlin's foes, but they rightly saw that radio gave certain messages a new resonance, that the medium empowered particular speakers. The question was, at whose expense? To Coughlin's critics, the answer appeared plain.

The experiences and reflections of some of Brinkley's listeners vividly illustrate radio's capacity for puppetry. Some of those who placed their faith in Brinkley ultimately came to feel that as listeners they sacrificed some of their autonomy to the voice on the air. They found themselves deluded by a man who converted his radio popularity into diamonds and yachts for himself, not social change or a medical revival for his followers. In fact, most who wrote to federal authorities objecting to Brinkley complained that his unregulated station drowned out other stations they wanted to hear; they attacked him as consumers who were denied their full range of choices. But a sizable minority of irate listeners instead rued a mass society in which the people did not count—or, more properly, counted only as objects to be manipulated. And they rued the mass medium that they found made that manipulation easier. By using modern technology, they complained, Brinkley gained an overwhelming edge in the con game they felt he was running. To those listeners, radio was dangerous because it enabled the few with access to the air to take advantage of the many who listened. The medium could devalue the individual, Geo Richards of Thurmond, West Virginia, sighed. Fooled by Brinkley's radio broadcast, he paid $750 to the doctor and received nothing but a laughable diagnosis, he said. The transaction convinced him that to those who controlled radio, listeners were not people, but sources of income. He wrote: "I am asking you as if this is what a person can expect by listening to the radio broadcast to be led into being robbed of wht is really all for the sake of having a radio to pass the lonesome days away and led to the chopping block as a beef is led to the slauter house to be killed while those that use the radio live in luxury at listers expence." Brinkley, in the 1920s, had helped pioneer radio as an advertising venue; in the 1930s Richards and others complained that through that medium the doctor gained an unreasonably persuasive voice. Richards, however, might be considered lucky: he lost only $750 to Brinkley's expert use of modern advertising technology. One listener, too ashamed of his plight to give his name, wrote to the Justice Department explaining that Brinkley's radio advertisements convinced him to seek out the doctor when he suffered a serious illness. The doctor deceived him, the listener lamented: "I told him on the operating table not to sterilize me. He went right on and done this to me just the same. . . . I would not have let him sterilize me for a million dollars. I was too young for this. . . . I am 30 yrs old and I did certainly not want this. He was too smoothe for me. . . . I feel like taking my own life over it at times."[40] To some of Brinkley's opponents, radio seemed a vehicle by which a smooth and far-off authority such as those Brinkley professed to hate could exploit the individual.

Like Brinkley's supporters, many of his foes worried that the rising influence of distant and impersonal powers would leave the individual impotent. But to critics of Brinkley—and Coughlin and Roosevelt as well—mass communication and those charismatic speakers who used it only exacerbated the problem.

Limited Amplitude

There was, as mentioned, a real tension here. True, a Roosevelt, a Coughlin, a Brinkley might use radio to sway their audiences. But it was also the case that all such speakers found limits to their persuasive abilities; the president, the priest, and the physician were generally at their most powerful when they did not try to lead their listeners places those audiences did not want to go. In other words, Roosevelt, Coughlin, and Brinkley found their power constrained by their audiences. Radio speakers could manipulate, but at the same time, such champions had the most influence when they actually fulfilled their promise of giving voice to their followers' concerns.

By the end of the 1930s, it was clear that even these expert radio heroes could not simply marshal listener support for any cause they chose, that they maintained their standing only as long as they remained in touch with their followers' thinking. Roosevelt in 1937 famously failed to persuade Congress to expand the size of the Supreme Court, despite taking his plea to the public. More tellingly, he tiptoed the United States toward involvement in World War II rather than leaping into the conflict. He did not dare go too far ahead of the audience he sought to lead.

For Coughlin and Brinkley, the limits of their radio appeals were starker. Both saw their popularity, their clout, and ultimately their public existences fade to black as the crisis of the Depression gave way to war. By the mid-to-late 1930s, Coughlin had increasing difficulty keeping his verbal hammer blows in time with the popular pulse. Emboldened by listener response to his radio addresses, in 1936 Coughlin broke decisively with Roosevelt and put forth a candidate, representative William Lemke from North Dakota, to challenge for the presidency. Coughlin might as well have run a representative from Pompeii, so deeply buried was his ticket by the Roosevelt landslide. Coughlin had pitted his popularity against Roosevelt's and lost badly. In the wake of that jolt and in the shadow Nazism and fascism, Coughlin's influence began to wane and his message shifted. He remained a popular radio speaker with political clout certainly; and to be sure, the fascist sympathies, bigotry, and anti-Semitism that increasingly dominated his talks had been part of his addresses since he first took to the national airwaves. But by the late 1930s more and more listeners and politicians could and did write him off as an extremist. His talks still drew huge audiences—though perhaps only a third of the tremendous listenership he boasted of earlier in the decade—but as he allied himself more and more closely with Nazi ideology the always-noisy complaints about him grew louder. As Coughlin and many of his listeners diverged, the priest's radio influence ebbed. By early 1941 he was off the air altogether.[41]

Brinkley found some similar limits to his radio influence: like Coughlin, neither the doctor's status as a radio champion nor his empire outlasted the Depression decade. Constant AMA pressure, mounting malpractice lawsuits, and increasing radio costs squeezed his finances. Other doctors, eager to cash in, attempted to copy Brinkley's methods, both creating competition for him and further discrediting his venture in his listeners' ears. And with war looming, Brinkley alienated additional listeners as he staked out an isolationist and increasingly pro-Nazi position—one that was not entirely altruistic. In 1937, for instance, he announced on the air that if one million mothers sent him one dollar each, he would keep the United States out of war; in doing so, he found a way to make isolation personally profitable. In 1938 he moved his hospital again, this time to Arkansas, but it was the last gasp of a dying icon, not a new beginning. After several years of financial scrambling and trying to hide his money from creditors, in January 1941 Brinkley filed for bankruptcy. Two months later, Mexico enforced the international radio agreement it had signed with the United States in 1940, a deal with provisions designed to silence Brinkley and speakers like him. The doctor was off the air for good. Without access to radio, Brinkley, like Coughlin, virtually disappeared.[42] The doctor's public status had been built out of air; he could overcome his opponents' attacks only as long as he had listeners and those listeners believed he spoke for them.

Regardless of such limits, there is an irony in both Coughlin's and Brinkley's popularity. Listeners who tuned their dials to either Brinkley or Coughlin, who funded either operation, did so in part because they responded to the speakers' attacks on powerful, impersonal organizations and vast, centralized institutions that curtailed individual autonomy. Those listeners sought representative heroes to help them navigate Depression-era America. But neither Coughlin nor Brinkley could entirely live up to his message in practice. As they fashioned their mass followings of the air, they moved closer to leading the very thing they claimed to despise. Empowered individuals, running their own lives through face-to-face transactions, had little place in either Coughlin's or Brinkley's organization. Radio made it possible for both speakers to rally people across large chunks of the United States, and to offer masses of people an alternative to accepting the bureaucratic structures of their society. Here radio gave millions of listeners a way to hear their frustrations voiced. But in terms of the structure of society, this was a rhetorical alternative only. Historian Alan Brinkley argues that Coughlin was doomed to fail because the decentralized world he campaigned for was already gone by the 1930s. Perhaps. In any case, the very means the priest and the doctor campaigned with—radio—pushed that world farther away.[43]

No doubt many listeners vaguely recognized the gap between Coughlin's and Brinkley's messages and their practices. For all their popularity, neither speaker ever won over more than a minority of Americans, and as World War II approached and their followers deserted them, both fell rapidly into obscurity.

Roosevelt's use of radio meshed more neatly with his agenda than either Coughlin's or Brinkley's, but the president's broadcasting contained a paradox too. Unlike Coughlin and Brinkley, Roosevelt did not set himself against a centralized and bureaucratic society. He implied that he would help accommodate the individual to that new world by personalizing it. And he used the radio to do just that. In the process, though, listeners' reactions to the president and their other radio champions complicated the meaning of radio democracy. Democracy, radio's champions and their supporters implicitly suggested, need not be about empowering people on a grass-roots level; the rule of the people might entail empowering particular far-away authorities. This would help change, or at least confuse, Americans' sense of where the will of the public could be discovered, of where one might hear the voice of the people. The people's speech might be found not in the mouths of the people, but in the people's mouthpiece—self-appointed or otherwise. Listeners widely cheered the radio president as a democratic hero—even as Roosevelt's supporters and opponents quite accurately noted that radio also helped him to amass tremendous personal power. He made a large, centralized government seem personally responsive to its citizens by further concentrating power and elevating an image produced for mass consumption over face-to-face intimacy.[44] Moreover, as radio enabled Roosevelt to build an unparalleled propaganda machine, it helped him tilt the balance of power in Washington from the legislature to the executive. Roosevelt and radio had changed the very nature of political authority in America—but had done so without forfeiting the president's democratic credentials in the minds of most listeners.

Perhaps Franklin Roosevelt was right, that Americans amidst the Depression were "ready to run after strange gods." Neither Coughlin nor Brinkley proved the prophets they claimed to be; radio, in fact, helped make both into millionaires more than messiahs. Roosevelt, of course, effected valued and far-reaching political changes, yet he too left a conflicted legacy as a popular champion.

Or perhaps it was not the gods who were so strange. By the early years of the Depression, many Americans were willing to overlook an inconsistency or two if they could find someone to help give them the means to deal with the challenges the modern world posed. Americans in the Depression were confused and scared, author Wallace Stegner remembered. "People," the writer recalled, "would listen to anyone who sounded as if he knew answers." To some, faced with problems they felt they could not solve on their own, radio voices seemed to offer those answers. Coughlin had, Stegner said in an often-quoted remark, "a voice made for promises." When their son's illness and their own financial precariousness overwhelmed W. L. Thickstrum and his wife, of Tullahoma, Tennessee, they wrote to Brinkley's station for guidance.

We "have serious problems which have not been solved in any of the usual ways, and are turning to any last, desperate, possible source of help," they explained later. For worried listeners, anyone willing and able to take up their causes was welcome.[45]

It was not simply the depressed economy that so frightened many listeners. Many of those who sought and those who shunned radio champions felt the webs of modern America spinning larger and larger, its strands confining their own lives and trapping them in a world perhaps too vast to influence. It was not simply those who latched on to some of radio's charismatic speakers who sought out ways to be heard in a national public sphere. As the two chapters that follow explain, academics and authors, in a very different fashion, also looked to the mass medium as a means of speaking to a mass society.

Radio and charismatic speakers on the air offered many men and women just that: a way to make themselves heard publicly and, perhaps, to make their interests count in the massive world of the twentieth century. As seemingly personal communication, broadcasting made it possible for listeners to perceive the speakers they heard as proxies. But proxies with forceful voices—as mass communication, radio enabled those speakers to reach, and perhaps sway, a national audience. It was precisely because broadcasting offered listeners something they desired so fiercely that radio champions were so dangerous. Huey Long, a highly controversial radio champion, declared that despite his autocratic power in his home state, his leadership was appropriate because he remained close to common Americans. "A man is not a dictator when he does the will of the people," Long said. "I have never been a dictator."[46] As long as audiences felt a connection with voices such as those of Roosevelt, Coughlin, and Brinkley, felt those voices had something to offer, listeners bestowed celebrity and power, often undue, on such speakers.

But part of the reason radio champions might be able to manipulate their audiences was because those proxies often did in fact offer those audiences something important to them. Many Americans craved ways to make their views heard across a national landscape. We must understand this impulse to comprehend the complexity of the concentration of authority in the age of broadcasting. Public intellectuals who criticized a mass-produced culture forecast that mass culture would shut down significant free speech and civic participation for most individuals. Many listeners experienced this viscerally. But for some, radio also made meaningful speech—albeit in an indirect and drastically altered state—seem possible in a mass society.

RADIO'S STUDENTS

Media Studies and the Possibility
of Mass Communication

Writing to a friend of his in 1938, Paul Lazarsfeld made sure to mention the big event in his home that July. By a long shot, he declared, his most important news was that he had gotten his radio repaired. On the downside, he facetiously admitted, his own productivity had "decreased fifty percent since."[1] As a listener, Lazarsfeld knew radio might hold the ears of its audience.

Lazarsfeld also believed in a correlated idea: that radio might offer select speakers a newfound voice, a voice that might reach and resonate with far-flung listeners. Since the previous year, Lazarsfeld's work, that work that he joked had declined so precipitously, had been the study of broadcasting in the United States. By the late 1930s, Lazarsfeld was the country's leading scholar of radio. As a student of the new medium, Lazarsfeld had learned enough to be deeply, deeply skeptical of radio achieving its promise. But he also saw in the new mass communication the hope of communicating meaningfully with the masses.

Two years before Lazarsfeld began his media studies, social psychologists Hadley Cantril and Gordon Allport declared that radio had created a "new mental world." By that Cantril and Allport meant that thanks to radio, listeners would forever understand the world in drastically new ways.[2] They were correct. But by the late 1930s radio had opened the door for another new mental world as well: over the course of the decade an expanding array of scholars made the effort to understand radio's meanings into a formal field of academic study. As a range of academics, generally from the social sciences, turned their attention to the burgeoning mass communications medium in the 1930s, they developed a new arena of inquiry.

And as they did so, as they explored and disputed the place of radio in Depression-era America, those scholars implicitly interrogated the idea of mass communication and a mass culture at large. To leaders of this new field like Lazarsfeld, radio had the potential to solve some of the problems of the twentieth century by opening up the possibility of communication with the whole of modern society. These social scientists noted that broadcasting failed to fulfill its promise, but they still believed that a mass medium might, if used properly, provide the communication needed for a huge and dispersed populace, knit together by invisible but real cords. A minority of scholars was still more eager to endorse radio in the United States, while another minority adhered to a more critical view, attacking America's mass culture altogether. On one level, these thinkers engaged in a pregnant fight over methodology: a battle over the nature of research needed to understand an immense world, over ways of knowing, over whether people should be understood first in terms of a social whole or as individuals. At the same time, students of the media clashed on questions of values: whether they prized education and high culture or capitalism, and what democracy should look like in the modern world. Most implicitly and most significantly of all, though, they laid out competing views of mass society and the nature of communication itself: was it possible to communicate with a mass audience in vital and valuable ways?

To most of the new radio scholars in the 1930s, mass communication held out the elusive possibility of keeping the modern world's mass public democratically informed. Not everyone agreed, but the majority of academics saw in broadcasting the possibility of a valuable, though volatile, voice in a mass society, a means of making essential communication possible in the twentieth century.

That this view so shaped academic understandings of radio in the 1930s was due in significant part to the work of Paul Lazarsfeld. In developing the new mental world of radio scholarship, no one played a more important role than Lazarsfeld. But if Lazarsfeld starred in the creation of a field, he was not alone on the stage. Lazarsfeld took a Rockefeller grant and support from first Princeton University and later Columbia University and parlayed that into the Office of Radio Research, an institution that defined and dominated the study of radio during the second half of the Depression decade. No American academic writing on radio during the Depression could entirely escape Lazarsfeld's influence. And yet, students of radio in the Depression occupied several distinct places on an uneven continuum of views.

Lazarsfeld and his followers at the Office of Radio Research and around the country sat in the middle. Two smaller groups resided on opposite sides, influenced by but apart from the weighty center. On one side of the median lay an array of scholars embodied by Herman Hettinger, a marketing professor at the University of Pennsylvania. On the other extreme, there was a member of the Office of Radio Research and Frankfurt scholar, Theodor Adorno, who actively defined himself against the Lazarsfeld school of thought. These three scholars did not comprise the whole field, of course, but they represent the key points on the continuum of competing ways those formally studying radio made sense of the media.

As researchers located themselves upon that continuum in the 1930s, the question of methodology loomed large. Pioneers in the field clashed over how to study the modern world, whether the fundamental entity to focus upon was society in aggregate or the individual. To those like Lazarsfeld and Hettinger, who believed meaning was constructed by the social whole, the exploration of radio in a mass society demanded scientific means to survey large groups and to assemble statistical data so the researcher could evaluate the medium's collective function. Such scholars were pragmatist social scientists, part of a decades-long search for new and scientific ways to investigate society in general.[3] To those like Adorno, who prioritized the individual and rejected any methodology or social system that did not, personal theoretical reflection upon radio could uncover the medium's meaning. The pragmatists measured the success of their research in functional terms. Dubbed "administrative research" by Lazarsfeld, this goal-oriented inquiry ultimately asked, how could society use radio most effectively? To Adorno, that was exactly the wrong kind of question to raise. Radio's faults were those of modern America's mass society, he argued; finding ways to use radio to support that society, then, made no sense. Embarking on what Lazarsfeld called "critical research," Adorno instead posed broad value-oriented questions, exposing connections between radio and a social system he saw eroding the individual.[4]

On the subject of social values, though, Lazarsfeld and Adorno had more in common than the two pragmatists. The mental world constructed by radio scholars was not a simple binary one. Lazarsfeld and Adorno each revered high culture and believed in the importance of a populace capable of thinking for itself. They measured radio by those standards, and both—though to wildly varying degrees—found the medium lacking. Hettinger, on the other hand, placed corporate interests and popular taste on a pedestal. Consequently, he easily endorsed the commercial system he observed. Both Lazarsfeld and Hettinger found promise in mass communication, but in significantly different respects. To Lazarsfeld and those we might call social pragmatists, enabling particular experts to reach the public with elevating messages could achieve their primary goal of bettering society. Hettinger, on the other hand, represented a commercial pragmatism: as he saw it, simply finding ways to give business an effective national voice would solve any problems.

For all three sets of thinkers, their visions of radio's social value grew from their visions of democracy in the modern world. Lazarsfeld and Adorno saw democracy in terms of a populace making choices for themselves. Hettinger, though, championed a capitalist version, one that united democracy with capitalism and consumer choice. But thinkers from the Lazarsfeld school and Adorno did not entirely agree here either. Lazarsfeld conceived of a system in which a mass public possessed enough information to make meaningful decisions within their social framework. To Adorno, democracy was rooted in free thought and thus took place on a more individual level. Clearly, Lazarsfeld and Hettinger disagreed sharply here, but both, unlike Adorno, accepted a mass society and thought in its terms.

At its heart, the intellectual gulf between media scholars articulated a division over the very nature of communication itself. Was communication possible within a mass society, or did the only positive ways of reaching audiences require individual connections? To thinkers like Lazarsfeld and Hettinger who accepted the arrival of a vast and centralized America, ways of communicating productively with that world would be vital. They saw potential, then, in a mass-produced and mass-consumed medium: it might enable particular speakers to reach the relevant listeners in modern America: the group. Their task, as they saw it, was to supply expertise to help make that system work. To critical theorists like Adorno who disdained America's capitalist mass society, though, there was little benefit and much risk in centralized correspondence with the whole. What truly counted was a direct and personal connection.[5]

As scholars in the Depression decade strove to determine what the new mass medium meant to America, three distinct views emerged. Among those academics seeking to define radio's meaning formally, Lazarsfeld's vision ruled the day. The leading thinkers shaping the new field of media studies generally displayed a commitment to the empirical study of social groups and the belief that mass communication might help address the challenges that the Depression and, more generally, the twentieth century, posed. In the Depression, most of the new students of radio saw potential in a media that might make it possible for a few voices to reach a large chunk of the public.

Paul Lazarsfeld and Social Pragmatism's Hope

When Paul Lazarsfeld agreed to head the organization that would become the Office of Radio Research, no particular love of, or even strong interest in, radio motivated him. Lazarsfeld's tremendously successful efforts to build an institution of social research reflected his attempts and the attempts by many social scientists to find a way to order their world in the Depression and early twentieth century. As Lazarsfeld came to understand that world, economic and political centralization and complexity had disrupted America, clashing with the basic ideal of an informed citizenry able to make many of the important decisions concerning their own lives. For those rooted in Lazarsfeld's thinking, reconciling mass society with their ideal of democracy required finding ways to give the ordinary people the information they needed in order to comprehend and function in their complicated modern world. Particular experts needed a means of speaking to, informing, and uplifting the national whole. Mass communication—radio—then might address the basic problem in a mass culture. Communication needed to be understood and practiced as a group, not an individual, activity. It could also be understood and accepted as a one-way process; radio could achieve its highest ends precisely because it enabled select speakers to reach a mass audience.

There was, however, no reason to assume that would automatically take place in a way that benefited society, Lazarsfeld and his fellows pointed out. They saw in radio the potential to benefit society by spreading socially valuable information and high culture. And they saw the task of the social researcher as offering an understanding of the medium's place in America that might help to fulfill that potential. That meant developing empirical methods of studying society as a whole: radio's meanings were contingent on a vast public's responses to the medium, Lazarsfeld asserted. Those who looked at radio as pragmatic scientists concerned foremost with the well-being of society had little doubt that radio fell short of their hopes for the medium. Their research findings and generally progressive values led them to critique the American broadcasting industry occasionally. But they did not rebuke the ideal of radio as mass communication. The social system in which the medium was used, Lazarsfeld argued, would ultimately determine broadcasting's value. But he and other leading radio scholars also implicitly accepted that the twentieth century demanded people be seen as a social whole. An appropriate voice that could reach that whole, they asserted, could make communication public—and meaningful—in a mass society and could invigorate communication and, in turn, democratic values for the modern age.

By the end of the Depression decade, what Lazarsfeld and his followers thought about radio dominated academic discussions of broadcasting. In the late 1930s Lazarsfeld developed more than just ideas about radio; he developed an institution. As radio became prominent in American life in the Depression decade, scholars across the country came to see a need to include it in their pantheon of studies. Consequently, in 1937 researchers

Hadley Cantril of Princeton University and Frank Stanton of CBS applied for and received a grant from the Rockefeller Foundation to set up a center for radio research at Princeton University. They tabbed Lazarsfeld to direct it. Lazarsfeld, an immigrant from Austria, first came to the United States in 1933 and spent two years traveling and studying the country. In 1935 he had secured a job at the University of Newark, where he ran a social science research center. By 1939 Lazarsfeld had secured a new grant from the Rockefeller Foundation, moved the project to Columbia University, and built his Office of Radio Research into the indisputable center for scholarly exploration of radio in the United States.

Lazarsfeld had mapped out the field of radio studies on his own terms. When the *Journal of Applied Psychology* published special issues devoted to radio in 1939 and 1940, not only was Lazarsfeld the guest editor, but by far the bulk of the articles came out of his radio research project; indeed, Lazarsfeld himself wrote so many of the pieces that he felt a need to adopt a pseudonym for some of them. Many of the era's scholars who were not directly affiliated with the Office of Radio Research walked the same intellectual path. Like Lazarsfeld, social scientists all over the nation focused on empirical listener research with a bias toward studying radio's use for education.[6] For the far-flung community of radio scholars, Lazarsfeld's office served as a sort of clearinghouse to guide and codify ideas from around the country.

Lazarsfeld's own focus on radio—and in turn the emphasis of his influential institution—evolved through the late 1930s. When he accepted the position as head of the radio project in 1937, he did so because he saw radio as an ideal arena in which to continue developing scientific methodologies for social research. Like so many scholars in the early twentieth century, Lazarsfeld was motivated by the prospect of finding empirical ways to study a complex society. As he wrote to Cantril when he accepted the directorship of the radio project, he had set up the research center at Newark for similar reasons. "I wanted to direct a great variety of studies so that I was sure that from year to year my methodological experience could increase—and that is, as you know, my main interest in research," Lazarsfeld wrote. For that interest, he continued, "I think that your project would do splendidly. Radio is a topic around which actually any kind of research methods can be tried out."[7]

By the time Lazarsfeld renewed the Rockefeller grant in 1939, he had built up a giant array of research methods that placed the Office of Radio Research at the head of the new field. But Lazarsfeld was also becoming more conscious of the need to put his methods to use, to show the Rockefeller Foundation that they were getting something for their money. Again like so many scholars of the day, Lazarsfeld and the foundation accepted the pragmatic notion that the measure of research's value lay in its practical application. And the applications Lazarsfeld focused upon reveal his sense that radio might serve as an instrument of progressive programming that could help all listeners cope with contemporary circumstances. A statement of the original goals of the project suggests that radio research was needed to insure that radio offered "the greatest good for the greatest number." However, the statement continued, before such studies could be undertaken scholars needed to develop research techniques and methods. In the application to renew the Rockefeller grant two years later, Lazarsfeld said a methodology was in place. Now, the proposal said, "it is planned to focus the work on the study of the conditions necessary to make socially significant programs more effective."[8]

Social pragmatists such as Lazarsfeld had no doubt as to what that greatest good was: providing uplifting information to foster democracy as they understood it. The real crisis of the twentieth century, even in the Depression, was one posed by an informational disconnect, Lazarsfeld believed. In an enormous but interconnected and centralized society, Americans increasingly felt confused by a world that stretched beyond their own vision and divorced from authority in their own lives. A mass society threatened meaningful autonomy, Lazarsfeld claimed:

> On the one hand, we cherish democratic traditions, which are based essentially on the idea that the citizenry themselves will make, and are competent to make, all the major decisions regarding life in the community. There is on the other hand this ever-increasing centralization of economic production, with the result that most issues which pertain to our jobs and hence the wellspring of our social existence are being decided by specialized technical or business experts far away from the scene of our activities.

The problem was not that those experts did not make good decisions; the problem was that ordinary Americans did now know about or understand the ideas such experts proposed. Without such information in the hands of the public, democracy was impossible. "Democracies," Lazarsfeld wrote, "can flourish only in soil nourished by the news which provides bases for free discussion."[9]

He and many of his colleagues believed in a democratic vision similar to the radio democracy growing in the 1930s, a democracy that valued information and the sense of public connection. Information, Lazarsfeld implied, could give people the ability to understand their surroundings so they could make some of their own choices in an interconnected world. On the flip side, information starvation, a byproduct of modern America's mass society, threatened the ideal of people maintaining some influence over their lives. Lazarsfeld wrote:

> The United States points with pride to its small and declining illiteracy rate. But at the same time science makes such rapid progress that the proportion of what a person does not know to what he knows is probably much greater nowadays than it was when very few knew how to write or to read. In fact, if literacy is defined as competence to understand the problems confronting us, there is ground for suggesting that we are becoming progressively illiterate today in handling life's options. And since it is no longer possible to make major decisions in local town meetings, the future of democracy depends upon whether we can find new ways for the formation and expression of public will without impairing our democratic form of government.

The problem of the twentieth century, Lazarsfeld might have said, is the problem of the line of communication. He bluntly stated as much in his proposal to renew the Office of Radio Research's grant: "Our chief national hazard is not the application of intelligence to our problems but the dragging lag in popular knowledge and acceptance that shackles the application of intelligence."[10]

Radio, properly employed, could help solve this problem. Most of those scholars devoting themselves to the study of radio believed that broadcasting could be used by particular authorities to overcome barriers to communicating with a mass populace and to give the full population tools to function successfully in their vast and interconnected modern society. Radio, Lazarsfeld and his followers felt, could be the organ that circulated vital ideas and information throughout America and in doing so pumped life into the democratic ideal of autonomous individuals. It could breathe an elevating high culture into the whole of the social body. Radio could be the sinew that held together and enlivened a national community. At their most optimistic, students of radio raved that for the first time there was a technology that could grant all of America access to the same information network. Radio's promise, Lazarsfeld claimed, lay in its potential "systematically to communicate the new social ideas which the immediate public interest so evidently requires a larger part of the population to grasp."[11]

Information broadcast by radio could instantly reach large segments of the population, leaping across geographic and economic bounds, many academics hoped. Americans once isolated from the information and understandings needed for modern life might be able to use radio to enable them to take their places in a national democracy. Sets were inexpensive and once listeners had a radio, they had the technology to access centralized information. Radio,

Hadley Cantril and Gordon Allport suggested, could carry "messages instantaneously and inexpensively to the farthest and most inaccessible regions of the earth"; it could penetrate "all manner of social, political, and economic barriers." By opening channels through which experts could distribute information to all of society, the Princeton and Harvard psychologists continued, radio served an "intensely democratic" function.[12]

To Cantril and Allport, radio might enable the finest professors to reach ordinary Americans flung across the country. This, they said, suggested "heretofore unknown educational opportunities." Such scholars focused much of their attention on radio's educational possibilities because they cared so deeply about finding ways to inform the public about their world. Lazarsfeld's office, for instance, often concerned itself with the question of education by radio and whether or not radio actually helped men and women who lacked other educational opportunities understand how to function in the modern world. The social pragmatists believed in education's potential to cure social ills. And they believed in radio's potential to aid in that cause. "For all its limitations," Cantril and Allport declared, "radio is without doubt the most outstanding innovation in the educational world since the creation of free public schools." Glenn Frank, president of the University of Wisconsin, where researchers studied the use of radio in the classroom, went them several degrees better and suggested that radio would be the biggest breakthrough in education since the printing press. Lazarsfeld would have dismissed such talk as groundless hyperbole, but he too considered radio tantalizing in this regard. Radio, he suggested, had the potential to make information accessible to men and women who, for instance, lacked the reading skills necessary to enable them to secure such knowledge elsewhere.[13]

When Lazarsfeld proposed to study "serious" programming, he meant not only talks about public matters such as news and political discussions, but also programs of classical music and the like. Radio could be valuable because it would give authorities of high culture a means to spread their cultural vision across the nation. This was no digression from the Office of Radio Research's broader social mission. Lazarsfeld and his colleagues retained a nineteenth-century understanding of culture: culture could be defined in terms of absolute standards, and high culture bettered the lives of those who experienced it. Lazarsfeld included classical music in his definition of serious broadcasting because he believed that listening to it could improve a listener's state as surely as any more material education. Researchers, then, could elevate society by finding ways to expand the audiences for high culture programs. "Helping to build audiences for serious programs could be, then, a new and effective form of the time-honored effort to raise the spiritual standards of the community," Lazarsfeld wrote.[14] If radio gave cultural experts a means to reach the mass audience—if broadcasting offered those experts voices in a public arena bubbling with cultural forms determined by popular tastes—the medium would provide a social good.

Ultimately, the social pragmatists admired radio's theoretical ability to help listeners overcome the sense of being lost amidst a world expanding and growing more complex. Radio could accomplish that, many scholars implied, not by enabling Americans to retreat into familiar local enclaves, but by embracing the national arena of life and helping listeners to connect to it. Successful participation in the whole counted for more than some abstract individual well-being. Consequently social pragmatists tended to find radio's promise of creating common communication exciting. It might unite the nation and turn simultaneously fragmented and interrelated millions into a mass community. For starters, radio could overcome individual isolation. Americans already belonged to a world wider than their horizons, Lazarsfeld suggested, but radio could help them to understand that. With nationwide broadcasts of news, events, drama, music, and talks, he said, radio helped give "the individual a new sense of participation in life beyond his own locale." Listeners could slip beyond the confines of class and spatial boundaries. Nationally instantaneous broadcasts created a personal sense of connection that no other existing medium could match. "What is heard on the air is transitory, as fleeting as time itself, and it therefore seems *real*," Cantril and Allport asserted. "It is not merely words and melodies that the listener craves. . . . When he turns his dial he wants to enter the stream of life as it is actually lived."[15]

By entering that common stream of life, these scholars claimed, Americans could begin to participate in a broad and diverse community. To Cantril and Allport, radio could serve as a social glue, breaking down the barriers of social stratification, offering all of America a common experience and creating "a consciousness of equality and of a community of interest." This was not a groundbreaking outlook in the 1930s: the first generation of academics to consider radio had also seen in broadcasting the hope of creating a common national community through national communication. But it was deeply important to the social pragmatists who formalized media study in the 1930s. Perhaps influenced in part by the rise of Nazism in Germany, they applauded a technology that they believed could provide a sense of commonality throughout the nation. "The Athenians gathering en masse at the Acropolis, had an ideal agency of unification. They could listen at once to their peerless leader, Pericles. Until radio was invented, America lacked an Acropolis," wrote Glenn Frank. "The radio is an agency of national unification whose development and freedom we must guard with jealous care."[16]

What radio could offer was not, it is worth making explicit, a democracy built around the idea of empowering citizens to speak to one another, of freedom for all to speak in a public sphere that encompassed the nation. These academics were concerned primarily not with participatory democracy, but with finding a way for experts to gain effective voices in their mass society. To this group, the problem of modern America was not centralization per se, but how to give the right authorities in central locations ways to connect with a populace spread across a continent—in other words, how to make the reality of a centralized culture a positive. Lazarsfeld stated as much in his proposal to renew the Office of Radio Research's grant. "We are living through a critical era of rapid, enforced cultural change in our American life," he wrote. "While science strives to discover and to formulate intelligent policies, it must also work at the problem of how to secure informed democratic consideration of and action upon such policies. The problem of securing needed social change is a crucial bottle-neck."[17]

Lazarsfeld's cohort believed that their research might help radio supply part of the answer. With such faith, Lazarsfeld and most of these radio scholars proved themselves pragmatic social scientists for the modern mass world. They hoped to use their research to help America better manage that society. To do so, they sought methods of understanding not radio in and of itself, but the responses of a huge populace to the medium.

These scholars who played such an instrumental role in constructing the field of media studies envisioned clear practical applications for their research. They identified a problem with their vast and complex society and sought to apply their study of radio to solving it. To Cantril and Allport, their research in the mid-1930s might help public officials "see that radio achieves its greatest social usefulness"; Lazarsfeld in the late 1930s explained that his office "planned to focus the work on the study of the conditions necessary to make socially significant programs more effective." This was an outlook true to the early twentieth century's intellectual orientation. Social sciences in general were taking a goal-oriented approach, Lazarsfeld wrote; the question framed by sociologist Robert Lynd—"Knowledge for what?"—loomed large in the minds of radio researchers as well. These were pragmatic thinkers, ones who defined knowledge in terms of its real world implications, not its abstract elegance. The success of a field of thought, Cantril and Allport asserted, could no longer be measured by the plausibility of its pronouncements or the number of textbooks produced: "The progress of social psychology must be determined, rather, by the incisiveness and validity of its analysis of significant social problems."[18]

The task for the student of radio, then, was not to uncover fundamental verities about the medium, but to develop the expertise needed to help others best use the instrument. Much as the scientific managers of previous decades had asked questions about how to use factory machines most efficiently, radio researchers would explore ways to use radio to achieve particular ends. Explaining his own scholarly orientation, Lazarsfeld wrote, "Behind the idea of such research is the notion that modern media of communication are tools handled by people or agencies for given purposes. The purpose may be to sell goods, or to raise the intellectual standards of the population, or to secure an understanding of governmental policies, but in all cases, to someone who uses a medium for something, it is the task of research to make the tool better known, and thus facilitate its use."[19]

Clearly, in suggesting that radio might serve such a range of purposes, Lazarsfeld dismissed the idea that radio had a particular inherent and immutable social and cultural meaning. These were thinkers in a post-Victorian tradition, arguing that meaning was constructed, not absolute; why seek to study capital *T* truths, if they do not as such exist? Radio, these academics believed, was a value-free device that reflected, not caused, modern life. Social developments and people, not technology, would determine radio's meaning, Lazarsfeld argued.[20]

By implication, the meaning of radio was embedded in the reactions of those who experienced it. In other words, this group of thinkers took mass communication's huge pool of listeners seriously. A program such as *Immigrants All—Americans All* might be designed to promote tolerance among native-born Americans by showing how different groups contributed to American culture, Lazarsfeld wrote. However, he continued, if the bulk of the listeners were immigrants, then the show could not possibly have its presumed effect. Radio's significance lay in those it reached.[21] To understand the meaning of radio then, a researcher needed to study who listened to what and why,. Consequently, the 1930s saw researchers all over the country taking their inquiries to listeners. More precisely, researchers often inquired into how large groups of listeners reacted to radio. If audiences held the key to radio's meaning, then collections of people—not specific ideas, technologies or individuals—were the essential social building block to study.

To Lazarsfeld, it was the challenge of figuring out how to conduct that sort of study that made heading the Office of Radio Research so appealing. If meaning existed in blocs of people rather than in fixed forms, neither rational abstract reasoning nor personal relationships alone could tease it out. "Social psychology," Cantril and Allport wrote in 1935, "can no longer pretend, as it once did, to solve the problems presented to it with some 'simple and sovereign' formula, or with a small handful of theories concerning the instincts or folkways of mankind. The science has passed beyond the stage of *a priorism.*" As heirs of pragmatism and social scientists and seeking to understand enormous social wholes, these researchers sought empirical methods of study; theories would take a backseat to statistical evidence. Cantril and Allport, for instance, set up a series of highly controlled laboratory experiments to test listener reactions to what they heard. And when Cantril, Lazarsfeld, and CBS's Frank Stanton launched their radio research project several years later, they placed that empirical focus front and center. The first step of the project, they said was to find workable scientific methods to use to study mass culture. In an era before a ready faith in polling, the question of how to understand the views of a gigantic and diverse public presented a real challenge. When the *Journal of Applied Psychology* ran two special issues devoted to radio, for example, the articles generally focused on technical matters such as how to run studies, handle data, use questionnaires, assess the representativeness of responses, and the like. Such matters—the task of coming up with scientific ways of investigating a mass society—drove Lazarsfeld. As a centrally run medium that reached a nearly national audience, radio made for an ideal laboratory for new kinds of pragmatic social research, he gushed. "Radio," he wrote in 1940, "has now made the whole nation an experimental station. A rather centrally controlled industry provides a variety of stimuli, the reactions to which can be studied and compared in all groups of the population."[22]

And by studying and comparing those reactions, Lazarsfeld and his fellow social pragmatists believed, they could learn how to use the medium effectively. Which returns us to the question of why conduct research: "Knowledge for what?" Lazarsfeld, in theory, believed research could and should be value-free, it should simply offer ways to solve existing problems. What ends to pursue and what problems to solve, Lazarsfeld wanted to say, were beyond the realm of the social scientist. His office, Lazarsfeld at times declared, was "essentially a service organization which has not to set goals but to help in selecting and achieving them." Everyone associated with radio was interested in employing it effectively to some end; the task of the researcher was to explain what made radio effective and let the policy makers decide what ends to put it toward. As scientific researchers, Lazarsfeld periodically suggested, students of radio should not allow their own value judgments to enter into their assessments; the usefulness of research was the final measure of its value.[23]

But, of course, Lazarsfeld and the group of radio researchers he represented did, in fact, measure radio by their own set of values. Lazarsfeld would be later criticized for endorsing the status quo by calling for value-free research, and his empirical methods provided a resource that the radio world employed for its own commercial purposes. But as much as Lazarsfeld suggested research do nothing more than apply scientific intelligence to help others use radio best, he and his colleagues had their own ideas of what "best" was. Radio had the potential to help experts distribute information needed to help the members of a far-flung public gain some measure of personal autonomy in America's mass culture. Radio, these scholars believed, was morally neutral, but with proper guidance, America might wield the new tool for society's benefit. The task of the researcher was to understand the tool's place in society and provide that guidance. That meant social pragmatists evaluated radio by these standards: did the medium serve as communication to salve modern America's wounds?

The answer Lazarsfeld reached was an unequivocal no. He and likeminded thinkers believed in the promise of mass communication, but after empirically measuring radio's impact, they had no doubt the reality of radio in the United States fell short of that promise. Just because broadcasting made spreading socially crucial information possible, the technology did not make that process inevitable. In order for radio to help bridge the information gap, Lazarsfeld's serious programming had to reach the masses of people. And that was precisely what the Office of Radio Research determined was not happening. In study after study, the office found that although the lower middle class listened to radio more than the wealthier classes, they did not listen to the programs that the Lazarsfeld cadre believed uplifting. Radio, for all its vaunted ability to reach the isolated, the nonreader, the poorly educated, could not elevate those listeners if they tuned in to *Jack Benny* instead of *America's Town Meeting of the Air*. Those "low cultural level" listeners—as Lazarsfeld called the economically and educationally disadvantaged—those whom he felt needed radio's potential boons the most, were, the Office of Radio Research found, the people benefiting least. "There is," Lazarsfeld wrote to the Rockefeller Foundation, "no foundation, for instance, for the vague optimistic hope that radio, as operating at present, will solve the problem of leveling up socially needed information among those not accustomed to getting such information through reading and other channels."[24]

Coming from someone who founded his optimistic assessment of mass communication on its potential "of leveling up socially needed information," this assessment damned radio bluntly. The Lazarsfeld cadre generally strove for a more neutral position, but such critical feelings occasionally leaked through. At times, the empirical researchers critiqued commercial control of America's broadcasting system—sounding, on the surface, just a bit like the radical James Rorty. Although radio had no intrinsic cultural meaning, social psychologists Hadley Cantril and Gordon Allport wrote, the medium's meaning was shaped by those who commanded it. "Like other instruments of power," Charles Siepmann wrote in a Lazarsfeld-edited journal, "radio is there to use and the outcome of its use depends on the integrity and purpose of those who control it." And since commercial interests ran the show, radio's meaning had been locked into those interests. William Robinson and Herta Herzog, both of the Office of Radio Research, found that listeners came to see buying radio-advertised products as part of the radio experience. And as radio linked broadcasting to advertising, listeners suffered. Instead of serving listeners, radio exploited them. As Cantril and Allport wrote, "Radio in America has room for improvement, and the improvement required is radical rather than meliorative. It has to do with the underlying profit motive which in broadcasting is, in the last analysis, not conducive to the *highest* standards of art, entertainment, education, child welfare, or even to basic freedom of speech. . . . In order best to serve the American public, radio should be removed from the dictatorship of private profits." At the very least, Lazarsfeld suggested, the corporate broadcasting system ensured that radio would serve conservative ends. Commercial stations and advertisers had a vested interest in preserving the status quo, and consequently produced programs that did so. Views that might challenge America's social structure rarely received a hearing on the popular airwaves.[25]

Lazarsfeld and his colleagues cannot, of course, be aligned with those who condemned radio in general or with those who criticized a mass-produced culture more specifically. Unlike the socialist Rorty, a fellow media scholar like the critical theorist Theodor Adorno—or even some radio voices such as Father Charles Coughlin—Lazarsfeld's cadre had no desire to break apart mass culture. The solution to radio's challenges lay not in decentralizing cultural control, but rather in organizing that control more fully and productively. Giving experts more influence, further developing social planning, not scattering influence, was the way to make radio serve its higher purpose. Reform needed not only to address radio, but to include the contexts in which listening took place as well.

To improve the value of radio, Lazarsfeld and his fellows believed, listeners needed contexts that welcomed serious programming. For instance, Lazarsfeld found that women who belonged to book clubs were more likely to listen to discussions of literature on the radio than those who did not belong. "The importance of the frame of reference is well known to educators," Lazarsfeld wrote. "The experienced teacher tries to give problems in arithmetic which grow out of the child's daily experience; he teaches history by starting with the local community." It was a listener's existing interests and affiliations that created the foundation that prepared him or her for serious programs, Lazarsfeld maintained. Similarly, Edward Suchman of the Office of Radio Research found that while radio could kindle a listener's interest in music into a flame, the ember of interest usually needed to exist before radio could fan it. A frame of reference must precede a broadcast, he wrote; radio could stimulate an interest to greater heights, but it could not create it from nothing.[26]

As Lazarsfeld admitted, stressing the importance of frame of reference was hardly a new idea in educational circles.[27] But applying that idea to education via mass broadcasting amounted to a call for substantial change. It meant reform could not be accomplished simply through measures internal to radio: improving educational programs or allotting more time for them, as many educators demanded, would not be enough. Instead, Lazarsfeld and like-minded academics suggested that to achieve the promise of radio, the position of the media in society needed changing. America, they hinted, needed to find ways to supply listeners with the frame of reference that would encourage them to seek out Lazarsfeld's serious programs. Schools, clubs, libraries, churches, even government agencies could be enlisted to stimulate listeners' interests in subjects that radio could then address. "Conditions that facilitate the acceptance of serious broadcasts need not be left to their 'natural' development," Lazarsfeld wrote. "Educators and social experts have never merely supplied material to feed interests already in existence; a strong element of promotion is necessary in any progressive program of action."[28] Reform would take a vast and interconnected effort; radio technology had bolstered the need for large-scale social planning. This was at once fairly critical of American broadcasting as practiced and fairly supportive of mass communication as a concept. Communication that enabled a central voice to provide programs to the diverse and far-flung populace could be socially valuable, but only with increased centralized planning and expert influence on society as a whole.

This endorsement of mass culture—even while occasionally offering real criticisms of radio in practice—opened the door for some softer critiques of broadcasting and for some to use this approach for ends entirely supportive of the existing system as well. Lazarsfeld, in fact, saw no problem in working hand in hand with researchers in the commercial radio industry on his studies. Indeed his office often benefited from network support. Moreover, social pragmatists could at times suggest that American radio needed only minor tweaking to achieve their ends. Better management of programs, not society, could bring about needed changes, this group sometimes said. Lazarsfeld himself occasionally admitted that educational broadcasters should study the techniques that made commercial programs so popular and apply them to educational ones.[29]

Such calls for better decisions on the part of centralized programmers without systemic changes—and the promise to use pragmatic social science to help programmers understand that decision-making process—were reforms that advocates of commercial radio could easily applaud too. Indeed, researchers like Herman Hettinger would adopt that very approach as they merged commercial values and pragmatic methodology.

That others with different values might share his approach to radio hardly disquieted Lazarsfeld. Making use of an empirical methodology was his primary faith. The only way to understand a vast society in the aggregate, he and like-minded researchers implied, was through a pragmatic and scientific approach. These thinkers believed in seeing people in terms of social groups and accepted the presence of a mass world. They were, however, guided by a desire to make that complex and centralized world compatible with their democratic ideal of people able to make decisions regarding their own lives. To Lazarsfeld and those who pioneered media studies with him, the answer to that challenge lay in building a modern mass democracy based on an informed, not a vocal or activist, public. As they articulated this vision, an elitism of sorts echoed through it. These scholars believed they knew best what kinds of information the populace needed and they relished a medium that might enable them to distribute broadly those materials. Theirs was, though, a vision in which radio served the needs of society as a whole. Lazarsfeld and others in his office and around the country seriously doubted that radio would do this. But they had no doubt that mass communication could do this. With appropriate and ample social planning, a technology that gave particular experts voices to reach the whole of society might ease the dislocations of the modern era. The cure for the ills of mass society just might be better and more, not less, use of their mass culture, or a piece of it—radio.

Herman Hettinger and Commercial Pragmatism's Faith

Researchers around the country gravitated to Paul Lazarsfeld's Office of Radio Research and social pragmatism, but not everyone who pioneered the formal effort to understand the media in the 1930s fell into that intellectual orbit. Radio research took place both under the auspices of ivory towers and on behalf of those who controlled broadcasting towers. Moreover, within the academy, a collection of scholars approached the study of radio as market researchers, committed both to an empirical methodology and to helping the radio industry realize its agenda. Herman Hettinger, who most articulately embodied such thinking, shared Lazarsfeld's hope that mass communication could amplify particular voices, enabling them to be better heard across the country, and, in turn, to improve society. But Hettinger's faith in the mass culture of the United States took another step as well: he embraced not only a mass culture in theory, but the institution of radio as structured in fact. Hettinger's confidence in America's centralized culture also included an endorsement of those who controlled the medium and in the value of a medium designed to appeal to a mass market.

Thinkers like Hettinger spliced the research vision of the pragmatist social scientist together with a commercial orientation focused foremost on what benefited corporate America. Clearly, such commercial pragmatists differed from social pragmatists like Lazarsfeld by defining the good first in terms of commercial interests, rather than considering society's interests as separate from corporate well-being. It would, however, be easy to overstate the differences between these two sets of thinkers. First, Lazarsfeld and others in his office periodically accepted an alliance between academic research and commercial radio. Second, the methodological overlap between the two sets of thinkers was not superficial: both believed in the importance of understanding the group. Third, and most important here, both sets of researchers applauded radio's potential to give particular people voices that could reach the mass audience. Both believed in the importance of speaking to the masses and imagined broadcasting could make meaningful mass communication real. Disagreement as to who should do the speaking, though, made for an essential difference between commercial and social pragmatism. Unlike Lazarsfeld, Hettinger equated corporate interests with the social good, and consequently lauded a communication system that amplified corporate voices. To Hettinger, commercial broadcasting and the United States more generally were exemplary and simply needed managerial guidance, not reorientation. At a basic level, the disagreements between the social and commercial pragmatists reflected diverging views of democracy; Hettinger envisioned democracy in distinctly capitalist terms. Ultimately, then, Hettinger went far beyond Lazarsfeld and his followers' hopes that mass communication might benefit the United States.

Almost as soon as radio was recognized as a mass medium, analysts sought out sophisticated market research to guide the medium. While the radio industry often drove that work, scattered scholars took the lead defining the field. By the early 1930s, Herman Hettinger, a professor of marketing at the University of Pennsylvania, had charted a path toward scientific, goal-oriented study of radio and its listeners. In 1930 the radio industry published its first ratings of various programs' popularity. That same year, long before many of the social pragmatists considered radio, Hettinger undertook a similar, though far more detailed project. His "Study of Habits and Preferences of Radio Listeners in Philadelphia" featured the same sort of empirical, listener-based research the Office of Radio Research would champion later in the decade. For radio to thrive, Hettinger said, broadcasters needed such systematic data about their field. "The groundwork of scientific advertising technique," Hettinger wrote in 1933, "is being laid in the work of agencies, networks, and individual stations, and in the infrequent research by universities and similar organizations." Over the course of the decade, that "infrequent research by universities" became more frequent. In the mid-1930s, for instance, the *Journal of Applied Psychology* published an array of articles by researchers seeking to study radio advertising empirically.[30] These early social scientists of radio accepted the industry's questions as their own.

Even beyond the cadre of academics who explicitly identified with business when they examined radio, the connection between the radio industry and the academics studying it proved strong. Commercial pragmatism frequently infiltrated the work of researchers such as Lazarsfeld and his followers. The Office of Radio Research, for instance, actively cooperated with the radio industry in many research projects. Frank Stanton, one of the research office's founders, worked as a CBS researcher in the 1930s and later served as president of the broadcasting company. Columbia University graduate student William Robinson's study of rural listening for the office was funded by the radio network. And in one of the project's reports written by H. M. Beville of NBC, Beville suggested that the directors of the research office hoped businesses would make use of the study. As head of the Office of Radio Research, Lazarsfeld promoted the academic-business link. His devotion to value-free research made it reasonable to provide research for anyone to use, and businesses provided him with a source of additional funding for the office.[31] The idea that business and public-minded organizations could be united to work for the good of society was not a foreign one in the 1930s: early in the New Deal several programs tried to bring business leaders into the public decision-making process, for instance.

Scholars like Lazarsfeld could periodically align themselves with the radio industry because from a methodological perspective, they drew no distinction between themselves and the work of market researchers such as Hettinger. When Lazarsfeld explained that his "administrative research" focused on how to use radio effectively, he consciously laid out a scheme that a huge range of interests could use. Researchers such as both Lazarsfeld and Hettinger fit easily under that umbrella. The two strains of thought accepted the day's common faith in scientific study to explain human behavior and to help manage the world. Hettinger and other commercial pragmatists, for example, engaged in detailed empirical studies of listener habits and responses to radio in order to help broadcasters better use the medium. In 1938, Hettinger coauthored what amounted to an empirically researched how-to book for radio advertisers. "Knowledge of [radio's] applications and technique has progressed to a point where guiding principles can be formulated regarding its scientific use," he wrote, explaining his work.[32]

Such an understanding, Hettinger argued, would enable broadcasters to use radio effectively. Like those who shared Lazarsfeld's mindset, Hettinger believed successful communication between a few speakers and a vast audience offered a potential boon. Hettinger, of course, had a very different idea from Lazarsfeld about who should be doing that communicating. To commercial pragmatists, their research served a vital function: helping broadcasters to make their advertisements effective. Put another way, such researchers saw their goal as enabling advertisers to make use of the tremendous reach of mass communication so they could contact and sell to vast audiences. Hettinger laid this mission out plainly in a 1933 call for scientific studies of radio. Three years later, Rensis Likert of New York University suggested that the real motive that underlay almost all the fledgling efforts to measure program popularity and to study radio was to help advertisers use the medium to their best advantage. "What we really want to know is not how many persons are listening to a program or can identify the sponsor," he wrote; "the real information that we desire is just how much influence the program in question is exerting on sales." The purpose of their work, Likert and others essentially asserted, was to help radio function as a more effective forum for advertisers.[33]

Commercial pragmatists placed the interests of corporate America foremost in their thinking because they believed serving those commercial interests benefited all of the United States. To Hettinger, giving select advertisers the ability to speak effectively to a mass audience served the general public. That—supporting the public interest—was the ultimate purpose of mass communication Hettinger asserted in his more reflective writing. "The advertising is merely the means by which [broadcasters] are enabled to render this service and, at the same time, to receive a reasonable return on their investment," Hettinger explained. Without that advertising, he also explained, radio as the United States had come to know it in the 1930s could not survive. Advertising unto itself was of limited importance; advertising as the essential means of sustaining America's entire system of broadcasting, though, was absolutely critical. "Broadcast advertising," Hettinger declared, "is the keystone of the so-called 'American System' of broadcasting."[34]

And to Hettinger, that broadcasting system was well worth preserving. He could imagine no other arrangement that better served the American public. Where Lazarsfeld—and William Orton and James Rorty for that matter—weighed the medium in terms of their ideas of democracy and found commercial radio at least somewhat lacking, Hettinger believed only commercial broadcasting made sense in his democratic-capitalist vision. The commercial system provided the funds that gave Americans the best and most popular radio programs, he said; capitalist competition gave rise to superior programming and kept government limitations on freedom of the press at bay. When Hettinger edited two issues of the *Annals of the American Academy of Political and Social Sciences* devoted to the study of radio, he prominently featured writings by some of its biggest cheerleaders, network and advertising executives.[35] For Hettinger, including such unabashed enthusiasts in his study made sense because he shared their faith in American radio.

Essentially voicing a mantra of capitalism, Hettinger asserted that the competition for listeners on the part of privately owned stations produced terrific programs. And because these were rivalries for large blocs of listeners, that meant the tastes of the many shaped programming. To Hettinger, a system of capitalist competition among broadcasters ensured popular rule of the airwaves and a victory for democracy: "The result has been a constant striving to win listener interest by stations and networks, with a consequent emphasis on the development of programs which will achieve this goal. Popular broadcasting accordingly has benefited. . . . It is a democratically controlled service, the broadcaster giving the public those programs which constant research and direct expression of opinion indicate to be the most popular."[36]

The existing system, Hettinger declared, democratically placed its faith in the market to sort out the interests of masses of customers. Financing radio through advertising, he said, kept the state out of radio, supporting a vision of democracy based on limited government. In Europe, he said, state control of radio had stifled the medium—and endangered free speech. "American broadcasting leads all of the nations in the extent to which it brings public issues before the people, and in the degree to which it provides a forum for a discussion of the questions of the moment," he wrote. Only a commercial system, Hettinger believed, could open the air to a wide range of perspectives. Capitalist broadcasting, he asserted, allowed for a variety of conflicting interest groups to be heard. If radio were narrowly controlled, he cautioned—ignoring the somewhat narrow commercial control that actually existed—the result could be homogeneous broadcasting. Democracy, Hettinger declared, depended on multiple interest groups having a chance to compete in the marketplace of opinions.[37] Any government intervention in that marketplace—any effort to regulate or shape programming—might threaten the freedom of the populace to vote with their dials and dollars for the ideas and entertainment they wanted.

Clearly, this vision of radio grew out of Hettinger's vision of modern democracy, one in which voters counted in groups in a market. In applauding a radio system that focused on listeners as large blocs and not as individuals, Hettinger, like Lazarsfeld and others, accepted the idea that in modern society people needed to be understood in aggregate. But the social pragmatists used that understanding to try to empower individuals in a small fashion—admittedly, a very small fashion. Hettinger, however, made it clear that communication in a democracy was not about giving the public tools to manage their world but about giving the public what was popular. Radio, he claimed, "is a democratically controlled service, the broadcaster giving the public those programs which constant research and direct expression of opinion indicate to be the most popular." Educational and news programs might often go by the wayside, he allowed, because in a democracy popular tastes should rule.[38] If commercial radio could support that, it could help sustain democracy amidst challenges of government activism at home and rising fascism abroad during the Depression.

Hettinger did not translate this—as scattered public intellectuals did in the 1930s—into a simple faith in some sort of rule of the masses. Hettinger envisioned democracy in terms of unleashing commercial competition and letting groups vie for control of the market. People—or, more properly, groups of people—counted chiefly as consumers with buying power, as members of a market. Recall, the commercial pragmatists were not interested in which programs were the most popular; they sought to know which programs produced the most sales. This meant majority taste would not necessarily rule; buying power would. Democratic radio, Hettinger explained, did not come down to appealing to the lowest common denominator in society. It meant, he said, appealing to the largest market—the group with the most spending clout. Radio, he said, therefore tended to aim at the middle class.[39] Broadcasters who succeeded did so, Hettinger asserted, not because they offered a uniform product that pleased a uniform mass audience, but because they balanced competing listener interests to attract one of the largest coalitions of desirable consumers. Democracy was not about majority rule, then. Free competition for social benefits—consumer spending in broadcasters' case—and, consequently, the ability to marshal social power defined Hettinger's ideal. To thinkers like Hettinger, yoking pragmatic research to businesses' pursuits of commercial gain made perfect sense: it meant mass communication would preserve their competitive democratic vision in the twentieth century.

The social researcher had a role to play in this capitalist democracy because modern mass culture demanded ways of applying the techniques of scientific management to the whole of society. Determining how a mass populace operated required specialized research, Hettinger believed. The radio industry needed scientific researchers to help it understand what choices large blocs of listeners were making, and how to take advantage of the amplified voice broadcasting offered. Businesses could not pursue simple laissez-faire capitalism in the twentieth century's complex social and economic system. Consequently, even as Hettinger vehemently rejected government oversight of radio, he called for bolstering the medium's central organization. He went so far as to suggest that networks and stations should form a loose alliance to centralize research for the benefit of all and to guide the industry internally. The demands of mass culture required the radio industry to centralize still further in order to enable broadcasters to reach audiences more effectively, he believed. The competition Hettinger so favored would benefit from pragmatic management. True, this managerial outlook might make the battles between broadcasters less vigorous, but it would benefit the businesses involved and, as Hettinger saw it, a mass society.[40]

Of course, in the Depression decade such faith in corporate America had been shaken. Amidst the economic crisis and New Deal, arguing the merits of capitalism untouched by government intervention took an unusual conviction. Similarly, Hettinger's commercial pragmatism was not the dominant scholarly view of radio by the mid-1930s. Most social scientists who defined the communications field over the course of the decade had little interest in following the lead of Hettinger and his cadre in bluntly adopting the radio industry's goals as their own. And yet, Hettinger's thinking resonated in an America struggling with the cultural changes of the twentieth century. Like Lazarsfeld and his intellectual kin, Hettinger believed in the importance of communicating with the mass audience to address the challenges posed by modern society. But Hettinger went beyond Lazarsfeld's hope that radio might be able to give central experts greater influence; the marketing professor had faith in the social merits of a broadcasting system that enabled commercial powers to expand their reach across a diverse and far-flung country. His ideas occupied a crucial spot within radio scholarship in the 1930s. Commercial pragmatists helped to shape the budding field of marketing and to pioneer the empirical listener-based research that lay at the heart of most academics' methodological approach to studying radio.[41] Moreover, even in the Depression, academics and big business often intertwined. Lazarsfeld and the social pragmatists could not escape—and sometimes welcomed—commercial influence. Ultimately, the social and commercial pragmatists both accepted the reality of their mass culture and the potential value of mass communication as a means of speaking within it.

Meanwhile, a critical theorist like Theodor Adorno at times virtually defined himself against it.

Theodor Adorno's Critical Theory:
A Considerably Less Charitable View

Herman Hettinger's vision of radio represented only one overlapping alternative to the social pragmatism that dominated academic studies of media in the 1930s. Working within Paul Lazarsfeld's Office of Radio Research, Theodor Adorno offered another. Like Lazarsfeld, Adorno considered radio's contribution to art, free thought, and democracy, and found the medium lacking. Unlike Lazarsfeld, though, Adorno envisioned no possibility that radio might benefit modern society. Where Lazarsfeld saw at least a glimmer of potential in mass-produced communication, Adorno saw horrors. To

Adorno, the idea that a form of mass culture, radio, could help assuage the pains of the modern mass world was absurd. Radio and American society both demeaned individuals, fracturing their connections to the real and important. That radio could do other than exacerbate this seemed ridiculous to Adorno because he focused his thinking on the means of production in society. Broadcasting, he asserted, offered only destructive mass-produced culture and communications, not the personal and direct ones he found so vital: even the concept of mass communication was an impossibility. To Adorno mass production itself—no matter who controlled that production—degraded culture.

Adorno's blanket condemnation of a mass-produced culture reflected a deep discomfort with the world of the twentieth century. He disdained what he saw as impersonal, homogenizing, and authoritarian strains in modern society and communications. Clearly this put the critical theorist in stark opposition to a researcher like Hettinger. Adorno's relationship to Lazarsfeld and his social pragmatism, though, was more complex—both personally and intellectually. At a basic level, both thinkers shared a common set of priorities: a devotion to high culture, and a sense that democracy depended on individuals making informed choices for themselves. These standards led both critical theorists and social pragmatists to look upon radio at least warily. But where Lazarsfeld adapted his values to meet the demands of a mass society, Adorno refused. He rejected the notion that modern cultural forces could be studied only in terms of a mass audience. Radio, by the very nature of the technology, had certain intrinsic—and negative—meanings that could best be understood through abstract thought, not through value-free empirical social research. To Adorno, any methodology that sought to make sense of radio within its cultural context inevitably overlooked the way mass production in communication and in society delivered related destructive effects. Radio, Adorno believed, inherently tarnished art; mass culture devalued personal thought, efficacy, and freedom. Although other early media scholars shared some of his ideas and over time those ideas would gain more attention, in the 1930s Adorno stood out as a rare media scholar who criticized not only radio in practice, but the very idea of communication and culture as a mass experience.

Although Adorno often deliberately clashed with the methodological approach of pragmatic researchers and although later media scholars would often look to Adorno as Paul Lazarsfeld's opposite, the critical theorist got his start analyzing American radio in collaboration with Lazarsfeld and his Office of Radio Research. Max Horkheimer, who worked for Columbia University, convinced Lazarsfeld to bring in Adorno to direct the research project's music study in 1938. Horkheimer, who had transplanted the Frankfurt

Institute's Marxist-influenced social criticism from Europe to the United States in 1934, wanted to bring his fellow Frankfurt scholar, Adorno, from Europe to the United States; Lazarsfeld wanted to add a critical thinker in the European tradition to the project. Interlacing the sophisticated theoretical concerns Adorno raised with an empirical methodology would create a valuable fabric of understanding, Lazarsfeld believed. In 1941 Lazarsfeld and Horkheimer collaborated on a special issue of the journal Horkheimer published to provide an example of how to weave together abstract theory and concrete research. Ultimately it turned out, though, that one loom could not handle both threads. During Adorno's three years with the Office of Radio Research he and other members of the project sparred repeatedly. In 1940 the Rockefeller Foundation, which funded the office, directed that Adorno be removed from the team. A year later he and Horkheimer moved to California and, over time, collaborated on several of the works that would come to express the Frankfurt School's critical take on mass culture.[42]

The critical theory Adorno and Horkheimer espoused showed up only rarely in the American academy during the Depression era and on the eve of World War II. And among the critical theorists, Adorno was the only vocal radio scholar. Where pragmatic researchers like Lazarsfeld focused on how to use radio effectively for particular goals, critical theorists sought to understand American culture at large. The administrative researchers wanted to build a better mousetrap; the critical theorists asked what was going on with our relationship with the mice. Such thinkers questioned the value of America's mass culture and commercial society, rather than focusing on how to operate efficiently within that structure. Over time, Adorno's understanding of radio would win a prominent place in the textbooks: as the relatively obscure critical theorists of the 1930s became better known in the United States as Frankfurt intellectuals in the 1950s and 1960s, many of them came to discuss mass communications and helped revise media studies. In the Depression, though, Adorno had bluntly positioned himself outside of the day's dominant scholarly approach to understanding radio.[43]

Many of Adorno's difficulties within the Office of Radio Research were no doubt due to his combative personal style, but clearly the friction rubbed more deeply. Lazarsfeld was in fact fairly tolerant of the theoretical queries Adorno posed. He had, after all, brought Adorno into the Office of Radio Research in part to broaden the empirical study of radio by asking new questions. On the question of methodology, however, the social scientist had no patience for the theorist. Lazarsfeld found Adorno's theorizing ungrounded and unscholarly. In a letter to Adorno, no doubt phrased as it was partly to soothe Adorno's ego, Lazarsfeld admitted that Adorno might well be right about some of the problems the American broadcasting system caused.

However, Lazarsfeld stressed, Adorno's complete disregard for scientific method fully discredited his points. To one who believed that radio's meanings played out in the context of a vast society, studies that did not seek to measure group reactions were suspect. "I have great objections against the way you present your ideas and against your disregard of evidence and systemic empirical research," Lazarsfeld wrote.[44]

Adorno did not deny Lazarsfeld's charges about his lack of empirical research; he virtually relished them. For Adorno, applying the scientific method to the study of art or society made little sense. Radio's import could be found in the technology itself and its means of production, not in terms of the changeable reactions of huge social groups. The task of the medium's students was not to survey the elements of a mass society, but to turn inward to consider the technology and the cultural system it inhabited. For instance, Adorno argued that studying the popularity of music told one little. Music, he said, contained its own meaning; to find that meaning a scholar needed to study the content and structure of the art itself. Consequently, in Adorno's lectures and reports on radio music, he sought to investigate radio music purely through the mind. Assessing radio, Adorno rejected the common pragmatic social science of the day and betrayed a rationalist streak: he believed certain inherent truths existed that could be understood by thought alone. Empirical research of listeners would ignore the true meanings of radio, Adorno believed. He considered radio and the American social order in the abstract, instead. And when the data others collected did not fit the ideas he developed, Adorno readily dismissed the data as "paradoxical."[45]

In part, of course, this was a dispute between Adorno and Lazarsfeld on how to know truth, a dispute that reflected the tensions between pragmatism and older approaches to social studies that had played out through intellectual circles since the late 1800s. Related to that, this clash spoke to the thinkers' disagreements on where meaning resided in modern society: in the social contexts occupied by a vast public—which, hence, required impersonal, scientific study to understand—or in abstract verities that might be individually accessed.

But in part, this was a dispute that demonstrated Adorno's commitment to particular social values—and his fear of research that claimed to be value-free. Exploring how to use radio efficiently and refusing to consider the value of that use struck Adorno as dangerous: social research that purported to be scientific and left out value judgments too easily played into destructive forces in the culture. His own approach to studying radio, Adorno admitted proudly, "certainly implies questions of 'value' . . . I would say that any analysis of social phenomena which has no notion whatsoever of what ought to be done and what is bad is completely meaningless."[46]

Lazarsfeld and the social pragmatists, of course, had a sense of what ought to be done and what was bad—despite their claims to the contrary. And to a real degree, Lazarsfeld shared his fellow European émigré's devotion to an elevating high culture and to preserving a degree of individual autonomy. Yet Adorno delivered a far more scathing critique of radio than the slightly critical but still-charitable social scientists that dominated the field. "Sometimes it appears as if the only content which offers itself readily to the new tool is that of destruction," the theorist said, speaking of radio in 1939. "We confess our utmost skepticism as far as the creation of so-called positive contents out of the tool is concerned."[47]

Quite simply, Adorno held out no hope for a mass-produced culture. Like Marxist public intellectuals James Rorty and Ruth Brindze, Adorno focused on the production of culture. Unlike those thinkers and Lazarsfeld, though, the critical theorist's foremost concern was not mass production's tendency to centralize control, but its capacity to give rise to endlessly repeated products. Adorno worried about what this would do to the individual's relationship to art and to individual thought. Radio, he argued, did not disseminate art and high culture, but destroyed them. Mass communication did not empower individuals, but emaciated thought and inspired an authoritarian society.

In his work on the study of music, Adorno swiftly concluded that reproducing good music on a mass scale was impossible. The mass production of music—or, more properly from Adorno's viewpoint, "mass reproduction"—automatically changed that music and drained its significance. It was not just that serious music might or might not be available; it could not be reproduced on the air at all. Even if radio could perfectly relay the sounds of a symphony—a point Adorno contested—something essential to a symphony was lost if a listener experienced it in a private room and not in its proper grand hall with other listeners. Just as a cathedral shrunken to the size of a shed lost its capacity for awe, Adorno said, so to did radio shrink and therefore transform a symphony. Radio listeners did not get even some watered-down version of high culture, Adorno said: "Our mythological farmer in the Middle West does not listen to a symphony in a mildly modified form, but to something fundamentally different, to something where the details make themselves felt much more strongly than the whole, while being distorted themselves, whereas it is almost impossible to get hold of the totality." That fundamentally different mass reproduction would rapidly do away with the original symphony, Adorno warned. The imitation, Adorno claimed, would drive out the real: listeners who came to know music through the radio would believe that the limited, disjointed radio music was the real thing.[48]

Moreover, Adorno continued, by endlessly imitating and reproducing the original instead of creating new art, radio transformed music from art to consumer good. Making copies of art, Adorno warned in language familiar to other Frankfurt scholars, fetishized the original work. The original would lose its intrinsic meaning as consumers, familiar with the copies, came to see the original as something to be prized only for its rarity, not its quality. "Paradoxical as it may sound," Adorno said, "radio in its present form, the technical means of mass reproduction, serves as a sort of romanticization of a symphony which is hardly compatible with its innate significance." Mass reproduction rendered the original art meaningless except as an item for consumption.[49]

In the end then, the effort to popularize art must dissipate its almost magical ennobling qualities—the very qualities that made it art. Radio offered Americans a means of mass-producing communication; that technology, coupled with the commercial culture of the United States, automatically meant that radio could succeed only if it won a popular audience. Mass production in the United States, in other words, demanded a degrading mass consumption. By making music into nothing more than a commodity, Adorno wrote, radio trivialized high culture. In the pursuit of profits, art would be dumbed down to suit mass tastes. No longer, he said, could it be offered as an ideal that might uplift listeners. Where once culture could be measured in terms of absolute standards, in the commercial mass culture that radio helped foster, art would be judged only by its popularity. Here again Adorno railed against the rising relativism of the early twentieth century: absolute truths and values, he feared, were falling before socially determined practicalities. And mass culture was, in part, to blame.[50]

The implications of this apparent assault on high culture and of mass communications in general were dire in Adorno's analysis. Radio undermined free thought, promoting an antidemocratic, authoritarian society. The problem, in part, stemmed from commercial demands. With its devotion to popular programming and maximizing audiences for financial gain, the radio industry treated listeners as nothing more than consumers to be manipulated for profit, not as independent-minded individuals, Adorno wrote in an internal Office of Radio Research report. Although not as broadly critical as Adorno, Charles Siepmann shared some of the theorist's leanings. Radio's great failing, Siepmann declared, lay in its overwhelming advertisements. Advertisements, he wrote, depended on large-scale manipulation and passive audiences to succeed. To the extent that radio was successful, Siepmann believed then, broadcasting and advertisements on the air in particular induced a "slave mentality." As critical theorists saw it, radio in the United States sought to reduce people's freedoms by directing them toward consumerist ends.[51]

Adorno's concerns, though, went beyond ones grounded in disdain for mass consumption. As Adorno saw it, a mass-produced communication undermined the very concept of communication itself. Real communication, he implied, took place on a personal level: it could forge connections between people or ideas; it fostered exchange, engagement, and thought. But broadcasting to the masses, he said, had the reverse effect. "Radio," Adorno declared, "is a prototype of vicarious communication."[52]

If genuine and personal communication could give one a sense of one's place, of understanding, of some control, vicarious mass communication might undo that. Radio, Adorno lamented, purported to connect the masses of listeners around the country with far-off events or people. In fact, though, radio offered not the event or speaker, but an illusion. Individuals did not actually take part in the experiences they heard; and, Adorno believed, the act of reproducing those experiences for an audience of millions warped the original meanings for those listening at home. To Charles Siepmann this false connection meant radio contributed to the individual's sense of inhabiting a vast world that he or she could not influence. The medium, he claimed, "constitutes yet another of these objective forces which have induced a sense of helplessness among individuals and robbed them of any real feeling of participation in events."[53]

Most insidious for Adorno, a mass-produced culture—from news and ideas to entertainment and art—threatened individual opinion and thought. The medium fed listeners a steady diet of "predigested goods." It minimized the space for free choice of what to hear and free interpretation of one's world. It trained listeners to accept the authority of the voice on the air; it taught listeners to prefer spoon-fed information to ideas they had to chew themselves. "The microphone," he railed, "does the listening for the listener." This meant radio might turn individuals into docile sheep rather than independent thinkers. The rise of Nazism had recently forced Adorno to flee his home country, but one did not have to witness events in Germany up close to reach his conclusion: mass broadcasting was inherently authoritarian.[54]

Adorno's critique was not simply a technological one, but one of capitalist mass society more generally. The dangers Adorno saw resided not only in the technology of radio, but also in the decision to yoke that technology to mass production. In other words, Adorno critiqued the structural workings of modern mass society and capitalism which he argued had been imposed upon radio, defining its meaning. He found radio self-contradictory: a technical advance, it served retrogressive purposes. The reason? The "conditions of production" in American society "chained" the thinking of those who shaped radio.[55] To Adorno and others who leaned toward the critical theorists' stance, radio's negative effects stemmed from and mirrored those produced by most facets of American life. Modern culture had created radio in its own image. When Adorno expressed his lack of faith in radio, he voiced his fears for his world more generally.

Because Adorno saw radio and the system in which it operated as essentially negative, it made no sense for him to follow the direction of Lazarsfeld and study radio from within that system. He had to get outside it, he believed, or he would fall prey to an insidious culture that operated with self-justifying values. Obviously, though, Adorno's disagreement with the pragmatists studying radio went beyond methodology. Adorno and Lazarsfeld—as opposed to a thinker like Hettinger—shared some basic social values and concerns about radio. Yet they disagreed in a fundamental but rarely stated way about the idea of communication itself. To Adorno, the very concept of mass-produced communication was self-contradictory. A mass medium could not convey that which truly mattered: high culture, genuine connections between people and ideas, and, at the core, space for independent thought. Real communication, Adorno might have said, was intrinsically individual and personal, not mass-produced. Mass communication, on the other hand, would ultimately corrupt not only communication but the individual as well. Where the pragmatist media scholars looked at a huge and interrelated modern world and wondered how to communicate with the people that made up that society as a whole—the masses—Adorno looked at the same society, condemned it, and seemed to dream about direct and personal communications that might defy that world.

Prior to World War II, of course, Adorno's work had a limited impact upon American media scholarship or thought in general. After the Second World War, though, these ideas would prove more influential.[56] In fact, though, even in the 1930s critical theory helped to frame the development of academic approaches to the media. The vision of the social pragmatists like Lazarsfeld dominated the emerging field, in part, because it occupied a middle ground, speaking to, drawing from, and distinguishing itself against both more critical and more commercial orientations. The work of someone like Herman Hettinger demonstrated that Lazarsfeld and the many like-minded researchers were not simply cheerleaders for the status quo.[57] Similarly, Adorno's thinking in the 1930s made plain just how distinctive the pragmatist social scientists' views were: the limits of Lazarsfeld's critique come most sharply into focus only in the light of another plausible contemporary reading of radio. It was not only Lazarsfeld that Adorno may have illuminated. Like other critical thinkers, Adorno implied that even as radio came to the forefront of American mass culture, there were multiple possible interpretations of that rise and of the possibility of mass communication.

To some extent, all of these academics studying the media carried on conversations that would not have squared with the experiences of most of radio's listeners. Adorno obviously did not concern himself with what the millions who heard radio thought of it. More than that, all three sets of thinkers— those represented by Lazarsfeld and Hettinger as well as Adorno—focused on radio as a medium to communicate with the masses. Most listeners did not dwell on radio's impersonal audience, but typically thought in terms of radio as a medium that engaged them individually.

In fact, the line between mass and personal communication, between public and private exchanges, was not so fixed in the radio age. The prospect of a particular kind of public voice tantalized many—both in and out of the academy. As we will see in the next chapter, to a group of ambitious writers who took to the air in the 1930s, radio offered the promise of energizing artistic and individual expressions by wedding them to public communications—and, as a result, giving them resonance in a mass society.

Clearly, not everyone shared Adorno's fear that mass communication quelled real communication. But in questioning the very nature of communication, he and his fellow students of radio asked fundamental questions while they shaped their discipline. Whether cheerleaders or critics of their age, these scholars saw a new world that raised particular challenges. To those who disdained modern America, mass communication threatened to substitute prefabricated group connections for genuine and personal communication. But most academics in the 1930s tacitly accepted the diminished importance of such direct exchanges in the modern age. To most of radio's students, the medium might—might—revitalize channels of communication precisely because it had such a public sweep: radio possibly offered a way to enable at least a select few speakers to reach the vast audiences of the twentieth century.

6

RADIO'S WRITERS

A Public Voice in the Modern World

Radio listeners who turned away from the popular *Jack Benny Program* to catch CBS's *Columbia Workshop* on a January evening in 1939 heard a powerful story of soldiers and civilians struggling with the carnage of the Spanish Civil War. William Merrick's radio play, *Forgot in the Rains*, demanded those who heard it take note of the devastation and human pain the war wrought. In insisting that listeners recognize and confront the traumatic public events shaking their world, the play issued a sharp challenge. Merrick's characters could not escape the anguish of the war. One of those characters, Manuel, in particular, had little tolerance for any who believed they could. The former soldier scoffed at what he saw as the insignificance of Picasso's paintings: "Dabble in blue and aquamarine—fret about line, effects, colors, philosophies, abstractions. Lovely stuff! Fine business, I say! But listen to me. When death sings in your ears and the only lines are flaming bullets, the only colors rotting flesh, the only effect, fear! Why talk of such trivial things as art?"[1] Merrick's question resonated through the United States during the Great Depression, on the eve of World War II and, more obliquely, in the twentieth century in general. Widely varying artists, critics, and others essentially wondered the same thing: was art indeed trivial in the modern world?

More proactively, a diverse array of artists turned the question around. How, they asked, could they make art matter in their world? How could individual artists project a voice that mattered

in modern America? For a select group of writers, dramatists, and directors, only a mass medium could fully meet that challenge. They saw in radio
a new artistic medium, a canvas for an entirely distinct form of creative
expression. But more than that, they saw in radio a stage upon which they
could discover a vital voice for a mass society. Particularly in the Depression decade, artists of all stripes—from painters and dramatists to novelists and musicians—often came to understand the core of art's importance
as public expression. A collection of radio authors eagerly engaged in that
project: they believed art needed both to address the populace at large and
to confront matters of political and social importance. But to this small
cadre of writer-director-technicians, the centralized modern world with its
vast and impersonally linked public seemed to allow few ways for individuals to make their voices heard. They concluded that if they conceptualized
their communication in mass terms, though, they might be able to meet
the challenges of their era. In modern America, they believed, they needed
a medium that could reach a mass audience. Radio, then, might make art
a social force and the artist's voice relevant. Through radio, these writers
believed they could answer Merrick's implicit challenge: radio could enable
the individual artist to speak publicly in a mass world.

With such ambitious goals, in the mid-1930s this cluster of innovative
writers began experimenting with radio as an art form. The Depression decade witnessed the collision of several cultural impulses—a drive to remake
the place of art in America, an effort to unite art and broadly left-leaning
politics, an evolving Modernist vision—and these radio playwrights were at
the center of that process. They believed deeply in public art, asserted a view
of the world that was both leftist-liberal and Modernist, and devoted themselves to exploring mass communication to convey their vision to the whole
of America. As broadcasting's programming took firm form, this collection
of writers—most prominently Archibald MacLeish, Norman Corwin, and
Arch Oboler—sought to push beyond common genres, envisioning radio as
a fresh canvas upon which they could repaint drama and poetry. They could
include daubs of sound on their palettes; and, through the airwaves, they
could hang their aural paintings in homes all across the nation.

That latter point was a crucial one to this group: it meant that radio could
offer the means to craft works that played a vital role in their society. As a
mass medium, radio could help make an individual's communication into a
public act—in a world that seemed to drown out significant individual expression. Only as public expression did speech or art have a living purpose,
these writers determined; and in modern America, they believed, that meant
their communication had to reach and move the masses. Radio's poets had
faith in the listeners who made up the mass audience—and wanted to find a

way to speak to that audience. Radio could both extend the reach of serious drama and poetry to grasp the vast population, and, practitioners believed, empower the public messages those words conveyed. To these radio writers, those messages were also critical to their agendas. They argued determinedly for the common humanity of all. Like so many in America's culture industries in the 1930s, these radio-wrights sought to merge their art with social-democratic advocacy at home and overseas. They appreciated the whole of the diverse populace because they understood the modern world as a place without fixed certainties, a place that could never be comprehensively comprehended, a place that could be accessed only through partial perspectives. With that in mind, they sought to take advantage of the medium of radio to capture multiple perspectives and to craft new ones to more fully understand the shifting reality they inhabited. This Modernism was not, these writers would have said, Merrick's fretting "about lines, effects, colors, philosophies, abstractions." It was integral to making sense of and functioning in the world of the twentieth century.

With this vision of public art, these writers came to see genuine communication as the mass communication broadcasting might make possible. The small collection of politically and artistically ambitious writers who turned to radio in the mid-1930s would have their optimism in mass culture blunted about a decade later, but for the moment, they found a magic in the medium.

Art of the Air

There is a linguistic challenge to understanding this effort to reimagine broadcasting. We lack a word for those who create aural performances. The writers who saw such promise in radio were not simply writers, poets, or dramatists in a straightforward sense. They set out to craft a new form of artistic expression, one that required a new kind of creativity, as film directors and playwrights differ from novelists. The radio play—as opposed to a play on the radio—demanded its creator write not simply in words, but in sounds as well. These radio-smiths had to imagine and layer together aural effects as well as language. They were not purely writers, but directors and technical experts as well. Perhaps better than most artists of the day, those who worked in radio would have sympathized with John Dos Passos's claim that the Modernist writer was a technician as surely as an electrical engineer was. Radio's writers—for lack of a better term—differed from other writers, then, in the basic stuff they used to compose their ideas and narratives. At the same time, though, these particular writers also stood apart from those who wrote scripts for the many genres that dominated the airwaves. Not everyone who wrote for radio self-consciously set out to create a new form of artistic expression. The divide was hardly rigid, but many focused more directly on telling stories or winning laughs through existing radio genres. It was a small cadre that set itself apart with a deliberate emphasis on innovative artistry and, crucially, on using that artistry to convey socially charged messages. These radio writers were a group that saw dramatically new possibilities of expression. Theirs was, radio writer Norman Corwin believed, a medium "to tickle the imaginings of all the poets in the world."[2]

Neither of the major radio networks, NBC or CBS, eagerly opened its airwaves to those writers who sought to turn radio into a theater for, simultaneously, artistic experimentation and civic commentary. Seeking to maximize their audience, network executives typically favored the broadly popular and shunned programming that they felt might affront listeners, particularly politically charged works. A medium that depended on advertising dollars and, in turn, a large-as-possible listenership, such executives might have said, could ill afford to take chances, with art or politics. And yet, late in the Depression and into World War II, to some there seemed the possibility that such a mass media could open up civic discussion and creativity, not close it down.

Prior to the mid-1930s few in radio talked about their medium itself as an art form. Radio programming was still young in the Depression decade. Before 1920 when Pittsburgh's KDKA pioneered scheduled broadcasting, radio was largely a scientific, not a programmatic, concern. And the shape of that programming did not coalesce until the end of that decade when the networks secured their hold on the airwaves, and innovative shows such as *Amos 'n' Andy* demonstrated radio's entertainment potential. How far and in what directions individual writers might take such forms, however, remained to be seen. Despite the spate of new genres of programs in the early 1930s, by the mid-1930s critics and those in the radio industry occasionally lamented the lack of experimentation on the air. "There is," wrote then-newspaper-critic Norman Corwin in 1935, "about as much creative genius in radio today as there is in a convention of plasterers and plumbers." That overstates the case: the invention of many of the on-air forms in the first place was, itself, a creative act. But the basic point—that once genres had been established they were often judged in terms of commercial popularity, not innovation, and then copied repeatedly—was valid.[3]

In a small but notable way, that began to change in the mid-1930s. A collection of writers interested in pushing the bounds of radio storytelling began finding outlets for their creativity on the airwaves. More than ever before, network radio welcomed these storytellers—welcomed them and put up the money for their programs. Since the late 1920s, when CBS and NBC worked out arrangements with local affiliates to carry network programs, the networks aired two varieties of shows, commercial and sustaining. Advertisers paid for and produced commercial programs, and, once they found popular forms, frequently stuck with them. The networks produced the sustaining programs. Although sustaining programs were supposedly one of the means by which the networks made good on the stipulation in their federal licenses that radio serve the public interest, in practice, such programs often served as fodder to enable affiliates to fill airtime not sold to advertisers. Not surprisingly then, sustaining programs were typically designed as low-budget productions. In the mid-1930s, though, that changed somewhat; the networks began emphasizing quality sustaining programming to a greater degree, and in some cases that meant giving radio writers freedom to make original productions.

Political pressures and, in turn, internetwork competition contributed to that shift. As Congress crafted the landmark Communications Act of 1934, debate centered on how much control to give educational and social service broadcasters. This might have been a moment that challenged commercial and national network control of the airwaves, and the networks sought to head off reform efforts. Ultimately, the federal government essentially maintained the status quo, deciding the matter in favor of commercial broadcasters. The dispute, however, led networks to take limited measures to stave off such controversies in the future: they hired public service directors; they bolstered their claims to provide educational offerings; they improved sustaining programming. Moreover, networks executives came to see sustaining programs as possible prestige builders. In 1936, for instance, NBC put together a grand program of symphonic music conducted by the renowned Arturo Toscanini. CBS too would offer its own elite musical programming. But bringing high-quality music to the air—seeing radio as a vehicle to convey existing high-culture forms—was not the same as opening up radio itself as a distinctly new art form. In the wake of the Communications Act, CBS also began exploring ways to take sustaining radio in entirely new creative directions. CBS president William Paley suggested in 1935 that unsold airtime might offer an opportunity for experimentation. As CBS's new programs began drawing critical acclaim, NBC followed suit. Such experimental shows never made up more than a tiny portion of a week's programming and they rarely received timeslots that would enable them to secure large audiences. But by the mid-1930s those who would become radio's leading artistically and politically minded writers had secured limited access to the air.[4]

That opportunity opened when CBS's pioneering experimental program, *Columbia Workshop*, debuted in July 1936. Headed by a studio engineer, Irving Reis, the show, as the name implied, served as a forum in which writer-directors could test new techniques in narrative, production, sound, and music. In the workshop, those creating a new sense of programming for radio were encouraged to innovate, to approach their projects as technicians as well as dramatists or poets. To Douglas Coulter, assistant director of broadcasts at CBS, the radio writers working on *Columbia Workshop* were scientists of sorts, performing "laboratory experimentation on the air." The program quickly won critical acclaim and paved the way for innovation in radio across the networks. It helped lift radio writers out of on-air obscurity, providing an arena to develop indigenous talent as well as attracting the efforts of established writers such as Archibald MacLeish and Stephen Vincent Benét. Perhaps most significantly, it gave the radio world *The Fall of the City*.[5]

No program did more to reveal the artistic potential of broadcasting— in terms of both stylistic innovation and social power—than MacLeish's groundbreaking aerial play, *The Fall of the City*. When it aired in April 1937 it convinced critics, writers, and many listeners that radio deserved serious creative consideration. "'The Fall of the City' proved this much," wrote critic and playwright Merrill Denison: "that radio can serve the dramatist who has something to say, and the ability to say it, as well as either the stage or screen." The impact of the play lay in the efforts of both MacLeish and director Irving Reis: each set out to craft a hugely ambitious radio play, not simply a play on the radio. Through the radio, MacLeish directly confronted some of the day's most compelling issues. His drama portrayed a nameless and universal city quaking at the approach of a nameless and enslaving conqueror. When the conqueror arrived, the citizens of the city rushed to subjugate themselves, casting off freedom and cheering, "The city of masterless men has found a master!" As the narrator told listeners, the masses preferred order to liberty; in the end, they bowed before an empty suit of armor. "The people invent their oppressors," the drama declared. Coming amidst the global rise of fascism, MacLeish's play carried an electric message. And it did so in electric fashion. MacLeish's play could have been produced nowhere else but on radio. The action unfolded as a live newscast. Listeners relied on a trusted voice, the announcer's, to bring the action to them. MacLeish had taken the very structure of radio and turned it into a narrative device—one that would prove enduring. Similarly, Reis's direction pioneered the art form, particularly in the realm of sound. To reproduce the noise of the masses in the city square where the play's action took place, Reis produced the show in an armory and recruited a crowd of extras. Reis and radio had turned sound into a vital actor.[6]

The performance wowed the radio world. Inspired by MacLeish, other artistically ambitious writers turned their attention to the medium. One critic suggested that with *The Fall of the City* MacLeish had created a literary precedent for radio. Another raved that MacLeish had given radio a dramatic purpose: "'The Fall of the City' proved to most listeners that the radio, which conveys only sound, is science's gift to poetry and poetic drama . . . that artistically radio is ready to come of age." Indeed, in the wake of MacLeish's play a flock of new writers began sending their work to network offices. "These men and women," claimed CBS broadcasting executive Douglas Coulter, exaggerating the access women had to the airwaves, "are beginning to see not only that radio is the most effective means of reaching the greatest number of people at one time but that it is a real art form, worthy of their best literary efforts."[7]

Among those new radio authors seeking to convey their messages, two swiftly took the lead: Arch Oboler and Norman Corwin. By the eve of World War II, Oboler and Corwin stood unquestioned as the medium's leading innovators and practitioners. Oboler had come to radio early in the 1930s, but it was only at the end of the decade that NBC loosened the reins on his ambitions. Oboler joined the network in 1934, writing for the horror series *Lights Out.* The series did not allow Oboler to tackle the political subjects that most interested him, but it did enable him to try new narrative techniques and offers a reminder that the line between seemingly pat programs and stylistically innovative ones was not always firmly fixed. With its midnight time slot, *Lights Out* afforded Oboler the opportunity to experiment with the relationship between narrative and sound, for instance. When, in one episode, a chicken heart grew so large it overwhelmed the world, Oboler expressed the organ's triumph by closing the play with only the sound of the heart's unceasing beat. In another program, Oboler turned a man inside-out, conveying the action through sound. By the end of the decade Oboler had helped establish the creative use of sound effects as an integral part of the radio play. At that point, after several years of fretting about the rising prestige of *Columbia Workshop*, NBC turned its extraordinarily creative late-night writer/director into a star: in 1939 NBC launched its own weekly experimental program, *Arch Oboler's Plays.* Though he would continue to battle with the network for more freedom to take political stands, for the first time Oboler had a forum for his socially charged, innovative, and often wild dramas.[8]

Late in the 1930s Corwin also produced a series named for him—first *Norman Corwin's Words without Music* and later *Twenty-six by Corwin.* Like Oboler, Corwin had helped convince his network that the success of radio would depend in part on the excellence of its authors. And as a radio writer Corwin had no superior. "He is to American radio what Marlowe was to the Elizabethan stage," wrote literary critic Carl Van Doren on the eve of World War II. "To Corwin belongs the credit for not only seeing what might be done with the radio script as an art form but also for doing it in a whole series of plays, poetic or humorous, which exhibit the full range of the art at present." From the start of his radio career Corwin had focused on the place of words in an aural arena. In 1930 Corwin, then working as a journalist in Massachusetts, volunteered to broadcast poetry creatively on a local station. Eight years later Corwin was doing much the same thing on a local New York station. Corwin's originality impressed CBS executives whose network sought such creativity. By the end of 1938 Corwin was directing the *Columbia Workshop* occasionally and producing, this time for a national audience, his energized poetry adaptations. His emphasis rapidly shifted toward verse plays, however, and by the end of the decade Corwin regularly directed his own artistically ambitious, increasingly topical, dramas and dramatic essays. That was, as he saw it, not simply a literary project in a traditional sense, but a new kind of creative and political challenge. "I would just like to see where I stand as a man and as a technician," wrote Corwin, explaining to a CBS executive his reasons for taking on a weekly program, "to take a fierce fling at the objective of doing the very best kind of radio I know how to do."[9]

Corwin and Oboler—along with many other radio writer-directors of the late 1930s—shared in their commitment to fashioning a new form out of their medium, and in their commitment to using their art for social purposes. Such writers believed in the possibilities of radio and specifically crafted their work for it. Literary critic Clifton Fadiman could have been describing Oboler when he wrote of Corwin, "From the very beginning he has conceived of radio as a special medium, not as an extension or transformation of the stage, the lecture platform, or the pulpit."[10] Not everyone agreed that the new writers had created a distinct type of art. Some English professors in the 1930s argued that radio works were too superficial to be classified as serious literature; and playwright Arthur Miller complained that when adapted for radio, a good story's wine "turns to water and its meat becomes fatty. . . . What I don't like about radio is the same thing I don't like about people who tell you everything twice." But for the most part, contemporary commentators felt this collection of radio writers had elevated their medium. In the mid-1930s, the Library of Congress's folk song archivist, Alan Lomax, had thought radio was "a pile of crap"; after hearing Corwin's programs, though, he "did a flip, I realized radio was a great art of the time." Certainly the writers on the air thought in lofty terms: Corwin, Oboler, and others fervently believed in taking radio to new heights, and they believed that on the brink of World War II they stood on the verge of finding a fantastically resonant artistic voice.[11]

Neither they nor radio, however, took that next step—at least not as they envisioned it. Laudatory historians have looked back on this brief period from the mid-1930s through the late-1940s as a time when the networks had a higher purpose, when they occasionally placed artistic quality ahead of simple commercial interests, when the air could hold some political stands. That goes a bit too far, even granting that this was an important moment of limited artistic freedom. CBS might have stated on its advertiser rate cards that the *Columbia Workshop* time slot was "withheld from sale," but since that slot was opposite NBC's enormously popular *Jack Benny Program*, the network had faint hopes of selling the period anyway. Moreover, radio activist-writers found their boldest efforts to craft political statements repeatedly blocked by their network overseers. As late as 1940, NBC executives refused to broadcast one of Oboler's plays, deeming it too anti-Nazi. Only when Oboler threatened to back out of his other commitments to the network did the play air. And an African American writer such as Langston Hughes routinely had his radio offerings rejected by the networks as too controversial. Moreover, the creative liberty that existed was short-lived. America's entry into World War II and the nationwide effort to rally on behalf of the country's struggle did open the air for certain kinds of political statements, but in the aftermath of the war, such opportunities chilled. The cold war and commercial concerns froze many bold and progressive writers out of the nation's airwaves.[12]

Still, in the late 1930s many writers found in radio the opportunity to undertake a new and ambitious project. For at least a moment, a collection of writers believed radio and mass communication offered them the power of public speech. For years, Oboler said, "I was pleading for the opportunity to do a weekly half-hour series of original plays written not for yeast, or soap or hand lotion, but idea plays written *for* the radio medium."[13] However briefly, he got the chance.

Public Speech, Public Art, and Mass Communication

Radio enticed writers like Oboler, in part, because they believed it could help the artist transform the nature of their communication to achieve what they considered art's high purpose: addressing crucial social issues in meaningful ways. They sought means of speaking publicly in modern America and believed broadcasting might hold the key.

Confronting public issues and political conditions was "the true ends of poetry," MacLeish claimed in a 1938 essay on the importance of public poetry. Romantics in the nineteenth century, MacLeish explained, had turned away from the public realm in their writing. That, he said, was a mistake contemporary writers sought to rectify. The ancient Greeks and Romans did not spend their time writing of "moons and maidens." They wrote about wars and governments and death and gods. "They had asked and answered the most penetrating questions of their time," MacLeish wrote. "They had known more of their time—and not only of its spirit but its economics and its politics—than those among whom they lived." Art, he asserted, needed to tackle such vital public subjects if it was to matter to people, if it was "to become again the thing it has been at its greatest reach."[14] And it needed to take those subjects to the public at large.

MacLeish was hardly the day's only writer concerned about the place of art and the artist's voice in American society of the Depression. Modernist artists had wondered for years how they could make art relevant to the majority of Americans. As reading rates declined in the early twentieth century, many found it increasingly easy to imagine reading poetry and fiction as little more than a quaint habit of a minority among a vast populace. When poet James Oppenheim helped found a Modernist journal, the *Seven Arts*, in 1916, he had hopes of making poetry integral to its era, he later wrote. In 1917 Oppenheim abandoned the project: the machine age, he lamented, "seems to have no real use for [the poet's] art."[15]

By the 1930s, though, more and more artists and others rejected that conclusion. Instead they sought new ways to spread their work through society and found resonance in confronting social problems. In the 1930s artists, museum curators, government officials, and others attempted to meld art and democracy, demystifying high culture and making artistic culture the province of ordinary people. Most famously, during the New Deal, the federal government helped take a lead in the effort to democratize art. The officials in charge of them believed programs like the Federal Art Project, Federal Theater Project, Federal Music Project and Federal Writers' Project could bring art to the millions. The Art Project employed some five thousand artists to paint murals in public buildings throughout the country and the Federal Theater Project's performances reached approximately thirty million people from 1935 to 1939. A director and actor like Orson Welles relished the opportunities this opened up: even outside of his radio broadcasting, he worked to bring classic dramas to wide audiences. And radio networks themselves voiced this value to some degree, declaring they served the public interest by widely disseminating high-quality music. For many, the Depression years expanded their ideal of the proper audience for the arts.[16]

At the same time, diverse artists and critics increasingly took on public themes in their work. They implicitly sought to resolve Oppenheim's dilemma by using their creations as political and social commentaries and arguments. Amidst the catastrophe of the Depression and the international threat posed by rising fascism, many artists eagerly confronted the political realities around them. The prominence of an array of leftist organizations in the 1930s, the effort on the part of those organizations to ally with New Deal liberals, and, most importantly, a rising spirit of general sympathy for progressive viewpoints all helped create an environment which engaged artists and cultural critics in a particular political sensibility. As historian Michael Denning has argued, the 1930s saw the rise of a broad social and cultural movement—the sensibility of the Popular Front—that included activist artists of all stripes, and which encouraged them—whether non-Communist leftists, Party members, or liberals—to merge their creative work with their social values. So a writer like John Steinbeck, a muralist like Diego Rivera, a film star like Will Rogers, and a singer like Woody Guthrie might all craft works honoring ordinary working people, solidarity, and cooperative values and suggesting social reforms. Critics and editors such as Malcolm Cowley, Edmund Wilson, and William Phillips too called for writers and others to address political themes in their work. Not all culture in the 1930s demonstrated left-leaning tendencies, to be sure. But by the mid-1930s circumstances and activism created among many artists an impulse to take political stances and to infuse their work with social-democratic and antifascist stands.[17]

These two meanings of public art—art for the public and art that took on public issues—did not have to go hand in hand. But the writers who would make radio their medium wholly embraced both impulses. Like MacLeish, most of these new authors of the air sought to reconceptualize art, to imagine the communication between artist and audience as communication between an individual and the public at large, not with only a select group. And like MacLeish, they understood the relevance of their speech in terms of social and political relevance. In articulating the need for public expression in both of these senses, though, these writers added an essential understanding. They understood the modern public in mass terms. And they embraced that mass public. Consequently, MacLeish, Corwin, Oboler, and others turned to radio. Only a mass medium could enable the writer to reach modern America's tremendous audience; and the airwaves could give those writers' messages tremendous power. Despite critiques that would later arise, then, in the 1930s to many it appeared mass communication would not damage art, or public communication more generally, but might revive both.

Art and speech, the radio writers asserted, gained value only by reaching others—and radio helped make that possible. In his radio drama *Appointment*, Corwin told the story of Vincent, a political prisoner, writing hymns to vengeance. His cellmate scoffed at him: "Has it [the hymn] more bars than your cell window? To which wall will you sing it when it's finished?" Without an audience even Vincent's song of rage had no meaning. Similarly, in his opening comments in the printed edition of *The Fall of the City*, MacLeish challenged writers to find new ways of reaching their audiences, to consider their audiences in the broadest possible terms. Books, MacLeish said, had isolated writers, permitting them to speak to only the few readers. The artist's voice needed to reach the many.[18]

Radio could give the writer a way to reach the far-flung audience that constituted the American public of the twentieth century. In a whimsical aural essay, *Seems Radio Is Here to Stay*, Corwin's narrator called to Beethoven beyond the grave to let him know that on radio his music reached millions of listeners in a single night. To Shakespeare, the narrator declared, "There's not a trace of mold about your poetry / . . . For the theater has grown / To take in all the stages of the land: / All villages and hamlets, / Cabins hard to get to, / Houses high on hills, and islands where the ferry plies but once a week. . . ." For too long, Corwin lamented, poetry had grown dusty, tucked away from the lives of the populace. "Radio, with the co-operation of poets themselves, will change all that and give poetry back to the people," he wrote. "Radio is the luckiest break that has come along for poetry in a long time." Corwin wrote his radio plays with this in mind: he deliberately set out to create readily accessible art, art that resonated clearly for the many. The connection with a mass public that radio fostered, Oboler marveled, made art meaningful: "The writer who has something to say and says it well can have, in a single half hour, a larger audience than Shakespeare had in a lifetime. . . . You write a play, the dialogue and sounds and music are transformed into electrical waves, and instantaneously what was only of you becomes part of a frighteningly huge and responsive audience. The satisfaction in that varies proportionately with what there was good in you that you put into the play."[19]

Oboler suggested that the satisfaction of mass communication varied depending on "what there was good" in the play because he, like so many among his cohort, believed in the importance of using art for a purpose. Oboler, Corwin, and their ilk sought to change their world, not just speak to it. That took moving the masses of people. To these writers, a lone individual speaking through conventional means had little means of effecting change in the mass society they inhabited. Only by reaching a huge audience and swaying the public as a whole could an idea gain currency in the modern world. To return to Corwin's allegorical play, *Appointment*, Vincent eventually escapes and makes plans to kill his former prison commandant. As Vincent readies a suicide attack from a hidden doorway, a fellow rebel, Mark, stops him. An individual act of vengeance is all but worthless, Mark explains: such a personal action would kill a man, but leave in place the evil system that caused Vincent and countless others to suffer. "Wait until there are more of us," says Mark, "in a bigger doorway." Corwin was not alone in his call for collective action. Two years earlier, Oboler had crafted a similar allegory. In his 1939 play, *Bathysphere*, Oboler portrayed a revolutionary doctor with a chance to kill his tyrannical ruler. In the end the doctor decides not to do so because the act would only bring about the rise of a new tyrant. The only way to win freedom, Oboler argued, is by inspiring the people to claim liberty for themselves. And of course, MacLeish's *The Fall of the City* made a related point in reverse: only because the collective citizenry chose to accept the nameless conqueror in an empty suit of armor does he have any power.[20]

Repeatedly, then, writers who turned to radio expressed the sense that individuals acting on their own could do little to reform the complex society of the modern world. The key to social and political change lay in bringing people together and rallying them to a cause. Other activist artists in the Depression shared that impulse, but radio's writers believed they had actually found a way to achieve these ends. By enabling the artist to reach a mass audience—the only source of real change—broadcasting could make it possible for a writer to better the world. In the modern United States, truly significant political expression required mass communication. "The radio audience is Mass in its best sense," wrote Corwin, "and if a man ever had a brief for freedom . . . the audience for *him* is the radio audience."[21]

Of course, these writers were not content with simply reaching a large audience; they wanted to move that audience. And radio, they believed, could enable them to do that better than any artistic form around. They applauded the new art radio made possible because the form itself bestowed striking power to their art. Radio, those writers who saw its artistic promise argued, made the artist's message supreme. Over the air, an artist's words themselves had an unmatched influence, in part because they created the whole of the play's world, and in part because they enlisted the listener directly in that act of creation. Movies and plays conveyed much of the story visually, Corwin wrote in an aptly titled essay, "The Sovereign Word." In radio, however, the word was almost all-powerful, he claimed. Corwin wrote: "There is no audience in the world which attaches more value to the spoken word than an audience of the blind—and a radio audience is just that. If a listener at home is moved to laugh or cry, to experience interest or suspense, the effect has been wrought by words alone. . . . If the word fails, the play fails." Like Corwin, MacLeish believed radio gave words a special, almost biblical importance. And that, he said, created an ideal bond between artist and audience. "There is only the spoken word," he wrote. "Nothing exists save as the word creates it." That, he continued, engaged the audiences' ears: "The ear is already half poet. It believes at once; creates and believes. It is the eye which is the realist. . . . The ear accepts; accepts and believes; accepts and creates. The ear is the poet's perfect audience—his only true audience. And it is radio and only radio which can give him public access to this perfect friend." Others agreed. Radio, they felt, enabled them to enlist the finest collaborator in the world to help convey their ideas: the imagination of the audience. The technology could give writers a direct connection to their listeners' minds; it could push listeners to internalize a writer's ideas. Through radio, Oboler believed, "the listener is living in his own mind the dramatic thought of the author."[22]

That unusual access to their audiences meant that, together, radio artists and their listeners could create an idealized and sharp picture of the world, so sharp it might serve as a path to action. In his play *Dark World*, Oboler suggested that it took imagining an alternative world in order to take steps to create it. Blind and paralyzed, Oboler's protagonist Carol experienced the world only through the stories others read to her. She was, of sorts, a permanent radio listener. But Carol did not lament what she missed. Instead of simply seeing the world with its flaws, she could envision its potential. "In my wonderful darkness—things as they *could* be—I had that," she said. "Things as they were? No—things as they *must* be." Radio dramas, the writer implied, enabled humanity to look beyond their society as it was to see something far more ideal. And unlike Carol, radio listeners could then step out of their idealized darknesses and together strive to make the world live up to the ideal. "Radio," Corwin wrote in reply to a letter from a Massachusetts prisoner who hoped broadcasting might be used to improve his conditions, "is the perfect medium for accomplishing not only penal reform, but a good many other repairs that our society seems to need."[23]

Clearly, to Corwin the world in the Depression and on the eve of World War II needed repair. As he and others in this collection of ambitious writers first began experimenting with their aerial medium, they felt a particular urgency to speak publicly because of the social and political turmoil they saw around them. They belonged to a generation of artists—broadly defined—that believed in opening a cultural front onto the political battles of their day. These were writers, thespians, musicians, singers, and more, all committed generally to ideals of social democracy, to antifascism, and to civil liberties. Whether or not particular artists voiced these views in explicit or polemical terms, such general values ran through much of the creative culture of the era. It was not coincidental that ambitious and civic-minded artistic radio bloomed in the mid-to-late 1930s, at the same time that the era's artists with progressive sympathies had their greatest influence upon American culture. Corwin, Oboler, MacLeish, and others in the genre belonged to the same movement. But few if any of the activist artists of the era spread their political vision as widely as those who put their pens to the airwaves. Unlike many of the better known politically conscious artists of the 1930s, radio's authors might count their audiences in the millions—even broadcasting less-popular sustaining programs.[24]

Like so many of their fellow cultural creators, those who turned to radio saw a multilayered threat: the Depression suggested that unbridled capitalism endangered liberty; the spread of fascist governments in Europe spawned hatred of others and threatened democracy and human life. Theirs was a civic call rooted in its age. The political timidity of the networks that controlled the airwaves muffled radio's artists' most vehement exhortations, but in their radio plays they still managed to advocate the defense of a set of clear, if general, values. To these writers, freedom appeared under assault, and they sought to explain what freedom meant in their time and to rouse their listeners to defend those ideals. They called for economic justice for all, democratic politics, and resistance to fascism. At the core of their advocacy, they shared a deep sense of the basic value of all people, common ground that underlay a pluralistic appreciation of difference. The mass audience was not simply a target audience for these activists, but, as a collection of ordinary people, represented the values they idealized.

To these public artists, their notion of freedom at times included economic liberty—but they equated that to security and equality, not the pursuit of profit. To many in the Depression, from liberal New Dealers to the unionizing working class to independent as well as organized leftists, unfettered capitalism endangered, not engendered, American ideals. Of course, critiques of centralized corporate capitalism could not be expected to fly readily through air controlled by the centralized corporations network radio represented. Still, in a testimony to the tremendous challenges to American values in this era and to this select group of writers' unusual abilities to create personal voices on the mass medium, they did mount indirect attacks. Oboler in particular railed against acquisitive individualism and declared the simple pursuit of wealth un-American. In *Mr. Whiskers*, for instance, Oboler portrayed an immigrant who feels he does not deserve U.S. citizenship after momentarily letting the desire for profit guide his actions. Duped into unknowingly selling marijuana, Oboler's immigrant shopkeeper declares that caring more about profits than people runs against the ideals of the United States. "To do things only for money is wrong anywhere in the world, but most of all in a democracy," the character explains.[25] The ruthless hunt for profits, Oboler indicated, produced a dehumanizing economic system; individuals needed economic security to have any meaningful freedom. In the first of his two plays titled *This Precious Freedom*, Oboler tells the story of a character who loses his job after thirty years at a factory; the company decides to fire experienced workers in order to increase revenue. Unemployment tears at Oboler's hero. "A man can't be a hero when he's out of a job," the character laments. Could he even be a man? the play asks. To Oboler and others, economic inequities lay at the root of all threats to human freedom. Returning to his play *Bathysphere*, the protagonist decides not to kill the tyrant after concluding that one man is not the source of tyranny: entrenched economic inequality is. Oboler's doctor concludes that gross economic division inspires political oppression as those with wealth seek to preserve their standing at the expense of others' freedoms.[26]

Many listeners, however, could have overlooked the economic meanings of freedom in the United States that *Bathysphere* championed: the driving allegory of the play attacked political threats to freedom and a tyrannical leader such as Americans saw gaining power in Germany in the 1930s. Indeed, at the tail end of the Depression, as World War II began in Europe, these writer-directors commonly defined their political visions in terms of international events. They continued to seek to define, preserve, and agitate for American values, but most often they explained them as civil rights set against the violent fascism in Europe: as democratic politics and free expression rooted in respect for the people and pluralism.

For many artistically ambitious writers, the most important thing the new medium for public speech could accomplish was to rally listeners to protect democracy against the Nazi threat. And as fascism grew abroad, networks proved more willing to air that message. For instance, late in 1940 a group of writers—including MacLeish, Stephen Vincent Benét, Sherwood Anderson, and Orson Welles among others—joined to form the Free Company. The group wrote a series of radio plays designed to remind America of its devotion to political freedom and individual rights and to inspire listeners to fight for those ideals.[27] Before that point NBC and CBS resisted taking sharp political stands, but artists worked their antifascist messages onto the air nonetheless. In his two acclaimed radio plays, *The Fall of the City* and *Air Raid*, MacLeish bluntly confronted clouds of war, asserting that liberty required a conscious defense and that new horrors were on the rise in Europe. Corwin earned his wings as a radio artist with his play *They Fly through the Air with the Greatest of Ease*, a searing attack on the inhumanity of the fascists in the war in Spain. Inspired by bombing raids on villages in the Spanish Civil War, both MacLeish's *Air Raid* and Corwin's *They Fly through the Air* scathingly told stories of innocent civilians annihilated by the impersonal cruelty of the war. To stress how dangerously fascists devalued human life, Corwin borrowed a line from Benito Mussolini's son, describing a plane gunning down refugees: "What a spread! Looks just like a budding rose unfolding."[28]

Repeatedly, Corwin, Oboler, and their compatriots, including the Free Company, issued calls to action. Obviously, in works such as Oboler's *Bathysphere* and MacLeish's *The Fall of the City*, the writers declared that common people needed to take a stand in order to resist dictatorship. And in *And Adam Begot* Oboler allegorically skewered governments that thought they could negotiate a peace with Hitler's Germany. NBC would not let Oboler weigh in directly on the 1938 Munich agreement appeasing Hitler, so he followed that settlement with a play about a nonviolent diplomat who, thrown back in time, attempts to reason with a threatening Neanderthal. In the end, though, the diplomat shoots the aggressor to save his own life. "You can't reason with unreason, can you?" the distraught protagonist concludes. "You've got to face force with force."[29]

Clearly, to Oboler events in Europe posed a real threat in the late 1930s. He sought to convince his listeners that the freedoms they so valued were at risk—even in the United States. Only by doing so, could he and other radio writers push audiences to defend what might otherwise seem abstract ideals. So in *The Man to Hate*, Oboler depicted the horrors of a state that punished those who thought differently, that trained individuals to think in step with the whole and to turn against those who did not. As audiences listened to a brainwashed child turn in his father, they heard of both the public and private crises posed by a regime prohibiting free thought. And in his second play entitled *This Precious Freedom*, Oboler assured his listeners, Nazism could reach the United States—with disastrous consequences: in the drama, only after Nazis overwhelm the complacent United States do Americans realize that freedom from political oppression matters to them. Corwin, though usually less blatant than Oboler, minced no words discussing what was endangered in the world of 1940. Integrating Daniel Webster's prophecy directly into his *The Oracle of Philadelphi*, Corwin cried for public action: "God grants liberty only to those who love it and are always ready to guard and defend it."[30]

Here two strains of thought behind the use of radio for civic art harmonized neatly. Radio's writers obviously cheered their medium because it enabled them to spur the mass action they believed was so crucial to effect public change in their age—radio could move a mass audience to defend liberty. But they also relished mass communication because communicating with the masses fit with the very core of the beliefs they championed. Ultimately, radio artists sought to address a mass audience because they had faith in ordinary people—a belief that regardless of differences, all shared in a fundamental humanity and everyone possessed basic rights and worth. Writing to one of his actresses, Corwin emphasized that aim even as he laid out his far-reaching artistic ambition: "I shall die a happy man if I have written ten lines that can move a stranger living a century hence, to feel the world more closely, or if I shall have contributed one jot of one little iota to the advancement of a universal brotherhood of man or to the establishment of a brave and clean and working peace." Obviously, any medium that could move the masses might also have dangerous effects—a point those who witnessed the rise of Nazism knew well. But to those activists who sought to create a new ethereal art form, their medium could elevate art and communication nonetheless because they genuinely believed in the value of those people who made up the masses and in the importance of speaking to them. We should remember, Corwin opined in his 1939 play, *Seems Radio Is Here to Stay*, that "even for man's monkeying with mania and murder, / He's still a noble article."[31]

Built upon this sense of universal humanity, radio writers' vision of American ideals led them to oppose both political systems that brutalized individuals and, more broadly, prejudices in general—stains that highlighted artificial differences between people. Set far in the future, Oboler's *The Laughing Man* featured a historian looking back on the first half of the twentieth century. In a painful commentary on the listeners' time, the historian could not stop laughing at what he read about a society that seemed preposterous to him. He could not believe the idea that people actually battled and went to war over matters of race. "They killed each other over a difference that they admitted didn't even exist!" the historian laughed and Oboler cried.[32] Diversity was nothing to rue. "Thank all the heavens and the gods for diff'rences!" Corwin wrote in *Seems Radio Is Here to Stay*. As Oboler expressed it, the democratic belief in an equal humanity, that everyone shared in human rights, was hardly radical—even if hard to find in practice. In *The Word* Oboler bluntly suggested rules to live by: given a chance to remake the world after God has wiped out most of humanity as punishment for the war in Europe, Oboler's protagonists strive to create a society without prejudice, one built on the idea that people are all, at some meaningful level, alike. All it will take, one character suggests, is remembering "to do unto others—as you would have them do unto you."[33]

If these seemed familiar principles—lifted perhaps from one of the Frank Capra films popular at the time if not from more hallowed documents—so much the better, the aerial artists would have said. They argued that their versions of democracy and freedom were bedrock American values. The very concept of America, Corwin suggested, meant including difference within a common whole, and recognizing the virtues of the variations. "America is all things to all her people," Corwin wrote in *Between Americans*: it could be the prairie of Nebraska or the coal mines of Pennsylvania; it could be the poetry of Walt Whitman or the poetry of the baseball diamond.[34] Each individual had a distinct identity, but in the value of that identity all shared something crucial. Their faith in that common humanity was what made the possibility of a mass audience appealing, not appalling, to this new breed of artists.

To Corwin, Oboler, and their cohort, an expansive idea of freedom and a democratic faith in the people lay at the core of American ideals, but they did not confuse those ideals with practice. The United States was not perfect, the writers often implied. Hardly. It did, however, offer the opportunity to strive toward a world without prejudice, brutality, political oppression, or economic deprivation. These writers repeatedly portrayed their version of America's ideals and, through their broadcasts, challenged Americans to reform their own country and live up to them.

Such occasionally transparent aims opened up these writers to the charge that they were creating works closer to propaganda than literature. Indeed, considered today, some of the weaker plays read more like tracts than dramas.[35] Others have more staying power; but even so, to dismiss the works on those grounds misses a crucial point. Non-radio writers in Depression era, from John Steinbeck to Mike Gold, were often charged with slipping into social commentary. In the face of political, economic, and social crisis, artists in the Depression repeatedly asked William Merrick's question: "Why talk of such trivial things as art?" Staring at a mass society that tended to mute individual voices, at the social and economic dislocations of the Depression, at the rising threat of political fascism and state-sponsored prejudice, a range of artists in the late 1930s often answered that question by seeking to change art itself so that it spoke directly to those not-so-trivial things. For instance, before he began producing radio plays, Orson Welles had already staged a distinctly antifascist adaptation of William Shakespeare's *Julius Caesar*.[36] To Arch Oboler, radio was an ideal medium for art, precisely because it made public communication possible in a mass society. His works were valuable precisely because they were topical. He wrote: "The dramatist who takes his themes from the headlines of his time because he feels strongly about today's injustices spends his talents and his energies without concern as to whether or not his works will play or read well the day after tomorrow; his one hope is that what he does will serve a useful purpose in this all-important today."[37]

Modernism on the Air

William Merrick's character, Manuel, would have no doubt agreed with Oboler on the need to "serve a useful purpose"; again, he considered fretting about "effects, colors, philosophies, abstractions" to be trivial. But to Oboler's cohort, matters of philosophy and politics, of creativity and civic-mindedness, of style and substance could not be so easily separated. The politically minded radio writers of the late-1930s sought to do more than express their values; they wanted to articulate their visions in new ways. Contemporary society, they believed, demanded that people change the very way they looked at the world in order to understand it. And with their works, these radio authors experimented with narrative forms designed to open up those new means of understanding. By offering audiences valuable ways to make sense of the confusing modern world, artists might make their voices meaningful within it.

MacLeish, Corwin, Oboler, and those engaged in their project were Modernists, with a vision of the world in which absolutes did not hold. Just as they valued the diversity of the populace and prized the importance of the people as a whole, they valued the diversity of ways of seeing the world and prized the importance of multiple perspectives. Reality, they argued, was complex, multifaceted, and changeable. Only by accepting that could they and their listeners find the means to comprehend that reality. Consequently, these writers wedded experimental storytelling techniques to their progressive politics. They played with perspective and fractured narrative; they explored ways to see a whole in terms of only fragmentary parts. Other Modernist artists had, of course, long since discarded the idea of a fixed view of the world. And those men and women, in turn, had often created a style that included ruptured narratives and many windows onto a subject. To radio's writers, the new medium seemed an ideal forum to pursue that project and to perfect that style. In striving to elevate their art form, these authors frequently discarded a consistent narrative voice and instead jumped from view to view. Stories leapt across space and time; to these authors, geographic and temporal distinctions needed to be bridged in order to understand the interconnected present. The writers emphasized sound to give listeners another angle from which to get a better hold on the whole of the action. Occasionally, the writers even took listeners inside the mind of a character, providing still another, deliberately limited, view of the story. In short, they rejected the single dramatic perspective of a traditional theater seat, creating instead works approaching the holograms of a later day—art that an audience might be able to see from many sides at once. True, writers producing scripts for sponsored programs also at times experimented with narrative. But for this particular cadre, that innovation directly served a political and cultural vision. Radio might enable these artists to communicate vitally in part because they could use it to help the populace adapt to a Modernist world. Without fixed absolutes, multiple views seemed necessary to gauge one's place in the world. At its core, their interest in the masses fit their Modernist vision: because all single perspectives were fragmentary, the group was essential.

Long before the Depression-era wave of radio artists took to the air, many intellectuals and artists had rejected the nineteenth-century ideal of a fully knowable world with fixed truths. The world the Modernists saw was unstable and knowledge of it was relative and partial at best. In the wake of Charles Darwin's *Origins of the Species*, thinkers had had to consider a world dominated by change, not constancy, philosopher John Dewey said. By the end of the nineteenth century this revolt against formalism was blooming. Intellectuals from Pragmatist philosopher Williams James to jurist Oliver Wendell Holmes attacked the a priori claims of rationalism; a writer such as Henry

James and a scientist like Albert Einstein questioned just how hard and fast the real actually was.[38] Following from this worldview came a Modernist artistic movement. The early decades of the twentieth century witnessed a burst of artists eager to experiment with perspectives and to break apart narrative structures. From expressionism and montage in films such as D. W. Griffith's stylistically pioneering features to the prismatic works of authors like Gertrude Stein and William Faulkner to the rule-defying comic strips of George Herriman, Modernist artists, in varying fashions, portrayed a relativistic and fluid world, often by examining it from multiple partial perspectives.

In the 1930s radio writers joined in that project, pushing it in new directions made possible by their medium. And they had no desire to practice in some cloistered cultural movement. Just as these artists thought in terms of spreading their political visions to the public at large, they hoped to democratize their creative visions as well.[39] These writers spread their aesthetic and ways of understanding the world to millions through the airwaves. To portray a changing and multifaceted world, they created narratives often told from many points of view. Those plural perspectives, radio's Modernists asserted, revealed more than any one of them could alone. Irving Reis, in his *Columbia Workshop* play *Meridian 7-1212*, made just such a case, showing the limitations of one person's view and giving his listeners an array of eyes. The play revolves around a reporter interviewing the director of a telephone time-reporting service for a story about the service. Standing in the reporter's shoes, the story is pointless—just the facts the director spews. But the play frequently leaves the reporter, traveling through the phone lines to peek at the people calling for the time. In their vignettes Reis created a compelling tale of the different meanings of time to different callers: from two London drunks arguing over the time in New York to a man waiting until midnight to commit suicide. Without those scattered perspectives, a listener would have gotten the same empty and meaningless view of the time the reporter found.[40] And, of course, no one person had access to all those points of view.

This was a point radio artists drove home through their frequent and distinctive use of narrators. "It is a fact seldom considered," Corwin wrote, "that radio has contributed the greatest stimulation to dramatic techniques since the invention of the movies. It has taken the ancient device of the narrator and developed it to points never reached before." In play after play, narrators led listeners through their stories. Yet those narrators rarely understood any more of the story than they could see from their own perspective: it was as if they were characters trying to figure things out along with their listeners. Emphasizing the incompleteness of any one point of view, Corwin frequently unsettled his narrators. In *Seems Radio Is Here to Stay* he dismissed one narrator and replaced him with another in midaction. Occasionally Corwin changed the narrator's voice by changing where it stood. In *They Fly through the Air with the Greatest of Ease* Corwin's narrator spends some time in a bomber with the flight crew, then he zooms ahead of the plane to visit the target town; later he flies along outside the plane, swoops inside, retires to a quiet corner of the sky to meditate on what he's seen, and returns to the plane. The narrator does not have an omniscient view of the bombing raid, but he reveals the perspectives of the flyers, the townsfolk, and a confused observer.[41]

The idea that reality could be known only through various lenses with particular tints, that even a narrator's knowledge might be partial, explains part of the appeal of a common trope in these radio plays: the narrator as radio announcer. When MacLeish told his story *The Fall of the City* as a newscast—that is, an event brought home by a radio announcer on the spot—he created a model that would intrigue radio artists for the remainder of the era. From Corwin to Orson Welles and the best-known piece of the day, *War of the Worlds*, radio artists often played with idea of the narrator as announcer. There were good reasons for the match. Announcer narrators, Corwin claimed, provided plays with an element of realism, simultaneously making a symbolic or fantastic story stand out and linking it to a listener's world. In the announcer, writer-directors had a trusted figure, one that Americans routinely turned to for information and guidance.[42]

At the same time, though, by casting the narrator as an announcer, these writers took advantage of that believability—and its limits. Even such a knowledgeable figure as the radio announcer—who had facts from around the globe at hand—could never directly and fully know the world, MacLeish implied. In *The Fall of the City* the announcer broadcasting the story at one point admits he cannot see a prophet leave the square: "She is gone. / We know because the crowd is closing. / All we can see is the crowd closing." MacLeish's announcer might be able to look at the multitudes in the crowd and convey what had happened, but he could not directly see the prophet's movement. And Orson Welles would later claim he intended his dramatic hoax, *War of the Worlds*, to serve as a reminder that even the trusted voices on the air needed to be considered critically: no one had a monopoly on the truth.[43] As narrators with particular positions in the story, the announcer characters demonstrated these Modernists' belief that the world could be known only through mediating viewpoints.

What made radio's writers particularly distinctive here was their work to wed that belief to the medium of radio itself. The periodic use of radio announcers was only one example. In their efforts to break apart narrative and to view stories through multiple lenses, radio artists often turned to means their technology suggested. These sound-smiths were not alone in the effort to integrate art and science. Some early Modernist painters, for instance, hoped to advance their art form by making it more scientific and mathematical. And early in the century Alfred Stieglitz attempted to connect technology and art when he asserted the artistic potential of photography.[44] But through the airwaves this melding often took place seamlessly and became a part of ordinary audience expectations.

The medium of radio leapt across space and time, for example, and artists readily adapted that technical attribute to their stories. Corwin admitted that he had "a sense of delight in the space-annihilating properties of broadcasting." That could be translated to drama, writers found. "The most flexible form in the history of artistic expression, radio offers great, exciting opportunities," Oboler hyperbolized. "There are no limitations of stage or movie set, there are no boundaries of time and space."[45] Space and time, radio Modernists essentially proclaimed, were not discrete or linear: only a medium that collected pieces of a common story across gulfs of miles or hours could make that story sensible.

By the late 1930s what would have been shocking barely a decade before had become common as newscasts regularly brought far-off events directly to listeners' living rooms. Artists took this practice and elevated it to new heights, bounding across imaginary spaces to bring listeners not only new views of the world around them, but occasionally even entirely new worlds. MacLeish, in *Air Raid*, did this quite self-consciously. In order to jar his listeners out of their parochial views of war overseas, he used radio to bring them right to its midst. His play began with a studio announcer telling the audience he would take them to a small town in Spain. Obviously, MacLeish never actually moved his audience anywhere, but by borrowing the very language of radio newscasts, he shifted his listeners' perspectives by thousands of miles. Corwin used radio's "space-annihilating properties" even more liberally. *The Plot to Overthrow Christmas* occasionally mimicked radio broadcasts as MacLeish's works had, but Corwin took his listeners to even harder-to-reach places. The audience follows the announcer to Hell, where a conference of historical villains selects Nero to kill Santa Claus; the broadcast then tags along with Nero on his trek to the North Pole, where he confronts St. Nick. Incidentally, the holiday is safe after all: Santa wins over Rome's famous fiddler by giving him a gift—a violin. Similarly, in *The Odyssey of Runyon Jones* Corwin sent his listeners on an unusual trip: in this case, all through the heavenly galaxy with Runyon looking for his dead dog's soul. And in *Anatomy of Sound* Corwin orchestrated an array of sound effects to create a forty-eight-second, nine-stop, around-the-world tour.[46]

But Corwin's masterwork in using radio to shift one's window onto the world was indisputably *Daybreak*. In this dramatic essay Corwin placed his audience on a plane traveling around the Earth at exactly the speed of the rising sun. By using radio to move spatially, Corwin froze time. Because Corwin's narrator kept pace with the Earth's rotation, listeners repeatedly experienced the same moment. Time itself proved contingent on where one stood. "A day grows older only when you stand and watch it coming at you," Corwin wrote. "Otherwise it is continuous." Listeners heard as morning dawned on a broad array of people around the globe, and in the process glimpsed connections across geographic barriers. "It [the sun] rises also now on two canals: the Erie in New York State, and the Panama. It's the same slip of morning to both ditches, though they lie two thousand miles apart," Corwin wrote.[47] By leaping across space through radio, Corwin had shown the relativity of even something as supposedly absolute as the sunrise. At the same time, he used his constantly shifting geographic perspective to reinforce his political vision: his inventive narrative uncovered the common circumstances of disparate peoples around the globe.

While *Daybreak* used the forum of radio to make time stand still, other radio plays used the medium to travel through time. On the radio, time did not have to follow the smooth, inexorable progression of a stick floating on a gentle current. Radio artists dipped into the stream at whichever points suited their purposes, pulling out snapshots from the past to help make sense of their present. Vic Knight's *Columbia Workshop* play, *Cartwheel*, dashed across time in an audio flipbook, for instance. Through a twenty-two-scene, fourteen-minute play Knight traced the history of a silver dollar across perhaps sixty years. Taken individually, each brief scene meant nothing; but by running together the montage of time frames, Knight created a cohesive history of the coin. It was Oboler, though, who fractured time most frequently. He often began a play in one period, then followed the narrator's memory and jumped into the past; in the process he turned the narrator into an actor, shifting narrative voice as he shifted times. Through several jumps to different times, the meaning of the present unfolded. For instance, Oboler began and ended his play *The Man to Hate* in the same moment, repeating the same line: "I will hate him, father! All my life I will hate him!" Without any time transpiring in the present, Oboler told the story of just what the line meant. Following the speaker's memories into the past, Oboler showed the moments that shaped the speaker's life and convinced him to turn his father over to a fascist state. Only when Oboler leapt to the moment when the speaker was told by his father, on the father's way to jail, to hate his betrayer did the play's present make sense. The present instant could only be understood by looking at it through many lenses, including those of the past, Oboler suggested. No moment, the Modernist radio writers and directors implied, could be fully understood outside of its historical context. Reality existed in fragmented parts; only by assembling them could humanity make sense of the vague, shifting whole.[48] Art, they essentially asserted, should bring the many distinct pieces together.

With this in mind, it should be no surprise that the radio writers of the late 1930s relished the use of sound. In sound, radio artists found another perspective through which they could explore their subjects. And by using sound extensively, they began integrating two often-separate artistic experiences: sound and story. This was, overwhelmingly, an innovation that can be traced to the 1930s and radio. According to critics, one of the keys that set MacLeish, Reis, Oboler, Corwin, and their ilk apart as radio artists was their creativity with sound. They composed their plays for the ear, using sound effects and music to tell stories, critics lauded.[49] In Corwin's *Appointment*, for example, much of the physical action was written in music and sound effects. One prisoner's failed attempt to escape was told entirely in music and a burst of gunfire; a successful escape appeared as music, the splashing of a swimmer and the chugging of a train. Additionally, sound made other perspectives possible: radio artists typically relied on music and sound effects to code their tracks when the story shifted venues. Listeners to Corwin's *Daybreak* could identify the motion of the narrator by sound: they heard first a storm at sea, then a foghorn, and later machinery and still later a volcano rise and then fade behind the horizon as the narrator kept pace with the rising sun. And Oboler moved his listeners into the past through audio effects. In *This Precious Freedom* listeners returned with the narrator to his days as a mill hand as the roar of machinery intruded on the present. Sound was, as Corwin expressed it, "more than mortar between the bricks"; it was part of the narrative.[50]

That meant sound was more than simply some stylistic adornment to these artists; by opening up the possibility of integrating sound into story, radio seemed to these writers to offer another perspective through which to understand the world. Because practitioners could write in sounds as well as words, they could better undertake their project of creating art that presented multiple views at once. Radio held out the possibility of telling a story from a perspective lacking in print. It is easy to imagine elite Modernist writers of the day welcoming an aural perspective to their works: John Dos Passos's newsreels in *USA*, complete with snatches of popular songs; William Faulkner's inarticulate character, Benjy, speaking in jumbled sounds as well as jumbled prose in *The Sound and the Fury*. Related aural styles were, in fact, seized upon in more popular arenas: in 1941 the film world discovered new possibilities for sound as a narrative device when Orson Welles brought his radio art to Hollywood with *Citizen Kane*.[51] Sound, the radio artists had found, could express another view of the many realities.

And that was a key reason radio's Modernists played with perspective in their works. They did not set out to provide multiple reference points from which the audience could triangulate some absolute truth. These writers applauded different but valid and coexisting views, and sought to make sense of them. This was obviously a narrative stand as well as an inclusive political one. Consequently these writers periodically experimented with an audio expressionism. Just as other artists did in literature and film, some of radio's writer-directors used sound effects and stream-of-consciousness narration to explore the individual's experience. In the first *Columbia Workshop* program to air, *Broadway Evening*, Leopold Proser tried to give listeners the sense of experiencing a walk through Manhattan. He did not focus on plot, character, or a concrete event, but on creating the impression the walk would have on a participant: he dotted his show with traffic, subways, and crowds, creating a noisy and occasionally incoherent program.[52] The effect might be incomplete and confusing, but it was intended as an honest expression of how one might experience the world.

Oboler did not originate the form, but it was he who took such radio expressionism to its greatest heights. He repeatedly told his stories through first-person participant narrators; the narrator-driven *film noir* of the following decade must have looked quite familiar to him. More than reporters describing the action, though, Oboler's narrators conveyed the story through their own partial impressions and jumbled thoughts. In his *Mr. Ginsburg*, when boxing manager Sam Ginsburg anguishes about his best fighter going blind, each of Sam's lines is punctuated by his fighter's plaintive cry, "Sam, why is it so dark?" Sam tries to dismiss it as simply a voice inside his head, but, like Sam, the audience heard the cry again and again. Fifteen times, "Sam, why is it so dark?" The voice in Sam's head felt painfully real—indeed, so real that the recurring cry convinced the once-cynical Sam to give up the hundred thousand dollars he earned from the bout that cost his boxer his sight. Frequently, instead of using dialogue, Oboler told his stories by tracing his characters' running thoughts, slipping from scene to scene as he followed the flow of a speaker's ideas. In *The Ugliest Man in the World* Oboler conveyed the blurred thoughts of his suicidal narrator in a blur: "gun in my hand gun in my hand in all my life I never had a gun in my hand smooth gun hard gun cold gun in my hand bullet won't be cold warm bullet hot bullet. . . ." In other cases, Oboler practiced his time travel by following the path of a narrator's memories, allowing the past to become the present as the speaker's awareness shifted.[53] Reality did not exist hard and fast, but in the slippery, flowing perceptions of the speaker.

In a case of life imitating art—or at least architecture imitating art—Oboler lived in a Frank Lloyd Wright–designed house with a stream running through his living room. The NBC publicity department delighted in pointing out that detail, but Oboler found it irritating. He rejected his reputation as a stream-of-consciousness writer. The idea that the style of some of his plays might overshadow their content upset him. Saying what he believed—not dabbling in stylistic issues for their own sake—most concerned Oboler.[54]

And yet, the ways he told his narratives mattered greatly to Oboler. Like that of his peers, Oboler's style counted not simply as an abstraction, but because it helped serve his purposes, helped him to express his view of the world and address its problems. His creative and his social visions were inseparable. Oboler's radio expressionism denied a fixed reality and simultaneously showed the limits and value of a single perspective. By making the relativity of the world plain and by opening up valid views of that world, radio Modernists like Oboler tried to help listeners better understand the surroundings they inhabited. Crucially, this cultural vision helped these writers implicitly mount their political arguments. All individuals had validity, they asserted. If no one possesses the complete truth, all perspectives are worthwhile and differences are valuable; only the populace as a whole can determine meaning; if a single omniscient viewpoint is impossible, dictatorial rule is inherently misguided. Communicating with the masses, they concluded, was therefore essential: in the Modernist world, mass communication was genuine.

In radio, writers found an ideal vehicle for their efforts to weigh varied perspectives. In their art they sought to create a sort of a prism in reverse. Instead of scattering reality into its pieces, they would take a scattered reality and bring its pieces together. That was, they suggested, the only way to understand the multifaceted world. And their radio work might help realize that vision. For radio's writers, radio served as a medium that could make sense of the changing world—and, in the process, make art matter in that world—in part because radio lent itself to the Modernists' stylistic project. And in turn, that project and their medium could enable them to pursue their political ends and, ultimately, their goal of crafting for themselves a public voice.

Muffled Voices

For those politically minded, Modernist writers who began practicing their craft on the air, then, radio in the 1930s seemed to hold out almost miraculous promise. In Corwin's *Descent of the Gods*, even deities marvel at the potency of modern technology, especially radio. Apollo, Venus, and Mars visit the United States and find themselves periodically humbled by scientific advances. Before leaving, Apollo addresses the country—by radio, naturally. "Man is greater than the gods in many ways, and lower than the ant in others," he announces. "He can fly the air and swim under the sea, and fling his voice around the world . . . [and] listen to the sounds of seven continents by fingering a knob."[55] Radio, it appeared, might be modern magic.

But if so, those writers who sought to employ that wizardry ultimately knew that they remained only the sorcerer's apprentices. No matter how much potential Corwin, Oboler, MacLeish, and others saw in broadcasting, they knew their political voices and their artistry could be constrained by the corporations who controlled the networks. Several weeks before *Descent of the Gods* aired, Corwin produced *Mary and the Fairy*. In that play, listeners had heard a fairy's magic regulated by the corporate world: the fairy employed by the Crinkly-Crunkly bread company to reward its loyal customers could grant wishes only according to the business's stipulations. Those restrictions meant that the play's protagonist, Mary, benefited little from the wishes she had won. Small wonder, Mary's fairy explained: the Crinkly-Crunkly company was not about to hand off its magic wand. "Don't you know that the only wishes that really matter are those you make come true yourself?" the fairy said. "Magic isn't very good because it's so unreliable." After all, Corwin commented in the notes for the published version of the play, "most of the accomplishments once reserved to genii have been taken over by industrial or political interests."[56]

Even the most acclaimed writers to take to the air had to contend with the corporate interests in control of their medium. Even at that moment when the airwaves seemed most open to left-leaning politics and Modernist experimentation, radio's writers repeatedly bumped against the demands of a commercial medium. The artists were interested in taking stands through narrative, communicating a pluralistic vision of the world. Network executives were interested in maximizing audiences for commercial gain, not in challenging norms. During much of his radio career, Oboler struggled to free his work from commercial control. He fought with NBC to produce socially significant works and he fought with his sponsors to keep commercials from intruding into his plays during the action. As radio tried to appeal to as many listeners as possible, Oboler lamented, it would not allow programs to take positions for fear of offending listeners. Radio, he wrote, deliberately calling to mind Neville Chamberlain at Munich, was in the business of appeasement. Obviously, artists who believed in addressing social issues would find their paints dried by such a policy. Corwin felt some of the same frustration. Broadcasters' obsessions with securing the largest possible audiences at the expense of other considerations seemed almost laughable to Corwin. In his whimsical *Radio Primer* Corwin attacked the radio industry's devotion to a system of rating shows by listenership. "If only one million hear you," he wrote, "you are talking to yourself." In the play a radio executive kills himself over a two-tenths of a point drop in the ratings. His partner sees a silver lining: publicity surrounding the suicide would no doubt boost listenership. This emphasis curtailed the kinds of programs networks would air in the best time slots—much to the frustration of the likes of Oboler and Corwin. The desire to do "good and ambitious things," Corwin wrote in a letter, was what separated humanity from animals—and, he added, hack writers as well. That desire, he continued, was incompatible with a devotion to listener ratings and commercial profit.[57] All of this meant that even in the late 1930s when the left-liberal impulse was strongest in America's cultural institutions, radio only allowed its writers to speak circumspectly.

But the climate of the moment did enable them to speak. The wave of optimism that this moment produced among these politically and artistically ambitious writers ebbed rapidly. World War II opened wide the airwaves to writers eager to give voice to antifascism: MacLeish headed a branch of the federal government's propaganda campaign and Corwin became the all-but-official radio poet laureate of the war. After the war, though, anticommunism and rising commercial pressures largely washed away hopes that broadcasting might make possible a civic-minded art. As was the case in so many culture industries, the cold war, rabid domestic anticommunism, and subsequent blacklists all limited the kinds of political messages and writers' work that could appear on the air. Since so many of those who envisioned radio as a distinct art form connected that vision to their belief in public art and their progressive politics, McCarthyism cut short their ambitious hopes for broadcasting. Formerly highly regarded writers and directors suddenly found the airwaves closed to them. Additionally, after World War II broadcasting became, to some degree, a victim of its commercial success and the changing cultural environment. In the postwar years sustaining programs fell by the wayside as advertisers bought more airtime: at the start of World War II only one-third of CBS's programs were sponsored; by the end of the war that figure reached two-thirds and was still climbing. Networks shed their less-commercial inclinations. Indeed, when the federal government broke up NBC, the network divested itself of its less-commercial wing. In 1948, when Corwin bumped into CBS president William Paley on a train, the man who a decade earlier had called for experimental programming told Corwin that radio needed to focus more on winning commercial approval and maximizing audiences. "You've done epic things that are appreciated by us and by a special audience, but couldn't you write for a broader audience?" Corwin remembered Paley saying. "We've simply got to face up to the fact that we're a commercial business." Whether Corwin's politics or his creative ambition concerned Paley, the effect was the same: the once-acclaimed radio poet was off the network soon after the encounter.[58]

The lesson here would have looked plain to those critics of radio who lamented a medium that concentrated control of communication into the hands of those few who owned radio's networks. Ultimately, politically minded, artistically ambitious writers like Norman Corwin and Arch Oboler could not express themselves fully on the air. To public intellectuals such as James Rorty and Ruth Brindze these artists' story would have affirmed their argument that a mass-produced culture limits who gets a voice. The point is valid and valuable: certainly, by the late 1940s, the once-golden dream of radio's aspiring artists had lost its luster.

But we need to recall that to many in the mid-to-late 1930s—and to others at other times—the matter looked much brighter. It was not wishful thinking that led MacLeish, Corwin, Oboler, and many others to seek out radio for their work; they saw real promise in the new medium. Precisely because radio was a mass medium, they believed broadcasting would enable them to speak meaningfully in their mass society. Radio, they believed, would make possible a new art form, one relevant in the modern world. Such writers and directors believed in public art—public art in two senses: in terms of reaching the general population and in terms of taking on civic issues. Only as public communication, they understood, could art matter: matter because radio writers had faith in common people and felt that only by moving the masses could an individual effect change in a mass society; matter because radio writers believed in the common human rights of those people and, in turn, in the importance of struggling for those rights. Their political and artistic choices reflected a public Modernism, one grounded in offering listeners ways of comprehending a multifaceted and relativistic world, not in stylistic digressions. "Poets will produce the greatest radio and radio will produce the greatest poets," wrote Corwin, expressing his sense of the excitement the medium held for art.[59]

By giving individual artists access to a vast public, radio could give their visions resonance. A minority of listeners, but a minority that numbered in the millions, heard and appreciated the works of radio's poets. It was not only these writers and an array of radio and literary critics who embraced the artists' view of radio. Orson Welles's 1938 production of *War of the Worlds* drew plenty of listener complaints, but it also attracted defenders, listeners who praised the show as a breakthrough. "Once in a blue moon, something comes to the radio that makes me glad deep down that I own one," wrote listener Skulda Baner, praising Welles to the Federal Communications Commission. "From morning to night radio is packed with Pacifiers. Give us who are weaned to more solid foods something to fill our bellies too!"[60] Radio, some listeners as well as a small group of writer-directors found, could give the artist a genuine public voice.

In fact, if this vision of radio as an arena for public artistic expression was not entirely successful, it was not entirely a failure either. Even as radio writers like Corwin and Oboler found themselves constrained by corporate control of the air, they helped demonstrate the possibilities of communication as a mass phenomenon. Broadcasting rarely served as a vehicle for ambitious or elite artistic expression in the 1930s or later, but it did push new narrative forms. Once-unsettling ways of telling stories would become at least somewhat familiar on radio and later television. More significantly, radio helped open up the question of who the proper audience for art should be. Whether or not the MacLeishes, the Corwins, the Obolers, the Welleses of the radio world produced brilliant works of timeless beauty was not the critical point. They were artistically ambitious, but part of that ambition lay in communicating with and creating an art for the public in a mass society. A crucial way to do so, they believed, lay in taking on socially important matters. Central control of broadcasting restricted them there, of course. But in the 1930s they found ways to be heard by millions nonetheless. In the process, they demonstrated the possibility that individuals could speak on the air and, consequently, powerfully amplify their voices to reach the modern public. The limits a mass-produced culture imposed upon artistic and political speech were real and, at times, very stark. But, as so often was the case with radio, here a group of radio writers found the mass medium could open up possibilities of new kinds of communication at the same time.

CONCLUSION

The year before Orson Welles and the Mercury Theater gave millions of listeners front-row seats for the fictional near-destruction of the world, radio aired another, smaller but much more real, crisis. And by many accounts, the medium performed more than admirably. In the winter of 1937 the Ohio River flooded cities and towns from Pittsburgh to the Mississippi River. In Louisville, Kentucky, where the situation was particularly dire, radio stations took an active role coordinating rescue efforts. Through the city's CBS and NBC stations, officials directed workers to danger sites and stranded residents, and guided those fleeing to safety and emergency resources. National broadcasts also confronted the chaos, drumming up volunteers and donations from around the country to help the drenched region. In response to a January 27 broadcast, over four hundred New York City volunteers raced to Louisville. To Anne Clatfelter of Attalla, Alabama, it made perfect sense to turn to radio in the tense situation: the medium could enable the whole nation to tackle the problem. "In this great time of flood, may I ask the NBC to be responsible for a Nation-wide prayer for the flood-suffers and for the rescue workers?" she wrote the network. "I am not in the flood district, but I want to help." In Louisville too, observers applauded radio's role. The *Louisville Courier-Journal* credited radio stations with making the evacuation of 230,000 people in three days possible. "Louisville's flood has been a means of demonstrating the underlying power of the people," the paper opined. "The catastrophe was radio's opportunity—and it has proved radio's triumph. Radio was the voice of authority."[1] To some considering the flood, radio seemed a way for humans to gain a measure of control in the face of a seemingly uncontrollable force.

This would prove true beyond the winter of 1937 as well. As Americans confronted their modern mass world, they used radio to help them find individual relevance in it: as many integrated broadcasting into their lives, they sought ways to maintain a measure of control within their own lives and they sought ways to speak out in the broader world. Listeners tuning in to both popular and political broadcasts used radio to personalize the impersonal wider world they felt around them: they engaged in relationships with the voices on the air, seemingly intimate ties that enabled those listeners to comprehend and feel directly connected to a vast public sphere. Radio's capacity for communication with the masses that composed that world suggested to some—particularly academics studying the medium and select radio writers—that some people could speak meaningfully to the far-flung modern public. At times, listeners merged the space radio opened to personalize the impersonal with the medium's potential to amplify voices, and felt—for better and for worse—that they might have a chance to be heard by proxy as powerful speakers took to the microphones.

There is a crucial measure of irony in all of this. At the same time that radio served as a means for millions of Americans to navigate their changing world, the medium was also a powerful agent pushing those changes. Centrally controlled, standardized for a vast and impersonally conceived audience, broadcasting represented and brought home the fundamental pieces of America's new mass culture. Public intellectuals of the day correctly implied that radio helped make a culture that was both mass-produced and designed for mass consumption into an everyday reality. On top of that, the ways many people interpreted and used radio reinforced the cultural change around them. Understandings and practices such as forming ethereal relationships, radio democracy, faith in particular speakers, social and commercial pragmatism, and the idea of radio as public art all implicitly endorsed the modern mass world.

We cannot ignore the oxymoron this represents. It raises an essential question: just how much control did radio actually offer the inhabitants of modern America? As the *Courier-Journal* suggested, the 1937 Louisville flood simultaneously demonstrated both the "power of the people," and the authority of radio. A little over a year later, in the *War of the Worlds* broadcast, Orson Welles implicitly asked whether people could remain empowered in a society in which radio held commanding authority. James Rorty, Ruth Brindze, William Orton, and other critics were right that broadcasting posed real challenges for individuals in America. The mass-production of culture and communication limits who gets to speak; a culture designed for mass consumption devalues individual distinction. As listeners and others came to terms with radio, those dangers persisted. The growth of mass communication raised the question of where ordinary people could speak meaningfully. Amplifying the voices of only the select few with access to broadcasting's towers narrowed cultural power, sometimes ominously. And by collapsing some of the divides between public and private relationships, broadcasting masked some of its dangers and helped change the nature of public participation. The sense of sharing a personal relationship with abstract and far-off entities might substitute for genuine interactions. These new obstacles to the exercise of individual authority undermined valuable ideals of democracy and civic participation.

Perhaps the most obvious place to say radio worked against the autonomy so many sought in the medium, was in accommodating Americans to their new mass society in the first place. For all that various Americans found their own meanings in radio—and they did so—broadcasting was not oppositional. Precisely because radio gave many a sense of being in some vague way empowered, it made certain kinds of standardization, centralized authority, and vast abstract connections more palatable. As radio helped usher in these changes, the medium strengthened the hold of America's mass society by enabling many to feel as though they had some control within it. That control, however, was sharply limited and could be deceptive. It did not extend to rejecting the fledgling modern world.

But that does not mean that the choices people made in determining radio's meanings did not matter. Far from it. Most Americans lived in a mass world by the end of the Depression. How they lived in that world—the decisions they made as they negotiated within the limits of their culture—mattered enormously to them. Individuals did not use radio to halt the rise of modern mass culture; indeed, broadcasting helped create that culture, with all its difficulties. But few of us can step outside of our surroundings entirely. We should value the choices Americans did make and the meanings they discovered in broadcasting as they sought ways to function within those surroundings. As Americans came to terms with the modern world, they found the role of the individual in society shaken. Broadcasting, however, gave many of them a way to personalize their mass world and, for some, to gain a voice within it. They used a piece of their mass culture to help them gain a sense of individual relevance in the world they inhabited. For those living with the cultural changes the twentieth century wrought, this was a real boon. We must recognize the limits—indeed the often unintended and dangerous consequences—of Americans' uses of radio. But we must also respect people's efforts to make their lives more livable, and consider the limited controls they did have. Americans could and did make some choices about radio, choices that helped shape their experiences in radio's America.

The meanings of radio, then, can be found in some delicate balances. On one hand, the arrival of radio and mass culture sharply limited individuals; on the other, radio proved beneficial for individuals as they negotiated within that mass culture. Both the costs and the benefits were real. Similarly, ordinary Americans were neither steamrolled flat by mass culture nor sitting at the controls. Just because radio's meanings lie closer to the fulcrums than the extremes, should not, however, imply that the arrival of radio did not matter or did not produce major changes. It did. Listeners' understandings of radio often recast the public in personal terms, blurring the bounds between the public and the private. As Americans considered the possibilities, and pitfalls, of a suddenly real means of mass communication, what communicating actually entailed came under question. And the United States debated over and redefined the meaning of democracy itself for the mass society of the twentieth century. Most of all, of course, in their understandings of broadcasting Americans' developed ways of living in the modern world—ideas and practices that are still relevant.

That world, the world that radio helped to bring about, persisted beyond the Depression and, indeed, beyond the decline of radio itself as the nation's dominant medium. The rest of the twentieth century made plain that America's mass culture continued to include trade-offs. Radio's America coalesced in the 1930s, but became still stronger in the decades that followed. Radio's heir, network television, took broadcasting to new heights, insuring that mass culture would increasingly structure the second half of the twentieth century. The tension between individual autonomy and the power of mass culture that Americans confronted in the days of NBC radio would remain current in the days of Disney/ABC/ESPN multimedia. And the risks that mass culture poses, risks that public intellectuals spotted in the 1930s, are still real and often underappreciated today. But the decades since the Depression have also showcased some positives amidst those problems. There can be an upside to the concept of a mass audience, a public that includes much of the nation; and having a way to speak to such a national audience can be invaluable. For instance, the civil rights movement of the 1950s and 1960s benefited from the possibility of airing racial discrimination to a national audience. By producing the same news for residents across the country, activists through the mass media could push a particular problem onto a national stage. By the end of the twentieth century, there were those who celebrated the idea of the national populace participating in a common culture and communication network. Democracy, political scholar Cass Sunstein provocatively argued in 2001, depends on inclusive public forums that make it possible for the general citizenry to share certain common experiences and civic information. Looking at the Internet and what he saw as the dangerous possibility of personalized communication filters, Sunstein applauded the idea of communication for a mass audience that William Orton abhorred almost seventy years before.[2]

Much about radio and the modern mass culture it represented should make us uncomfortable. But condemning—or, alternatively, cheering—it is too simple. On a 1933 broadcast, humorist and social commentator Will Rogers pointed out the damage radio wrought; and yet, ultimately, he could not ignore the millions of listeners who found the medium so welcome. Radio increased the power of a centralized voice, he began: "This rat-trap contraption that I am talking in front of here is now fixed up so that any of us . . . can bore the whole world in one evening." It brought a wider world to listeners' and nations' doorsteps, but that could cause problems, he lamented: "They tell you [radio] has been a great cementer of good will among countries. I don't suppose there was ever a time in our history when as many nations were just ready to start shooting at each other as they are now. Nothing that makes people acquainted makes friends." And, he continued, the medium's far-off and abstract commercial sponsors constantly told listeners how to behave: "At any hour of the day or night, tune in and someone is telling you how to live, how to vote, how to drink, how to think, when to wash your teeth, when to wash your hair, when to cut your whiskers, when to see your doctor, and how to see your doctor." But, he concluded, people still found radio made their lives better—and to Rogers that redeemed the medium. What people thought and felt mattered deeply to him. Radio might have helped create the jarring modern world, but, Rogers believed, anything that people could use to make that world more livable deserved some credit.

> Honest, no other nation in the world would stand for such advice as that. But we do, and we like it. So, the only thing that can make us give up our radio—we will never give up our radio—unless it is poverty. The old radio is the last thing moved out of the house when the sheriff comes in. It is an invention that has knocked nobody out of work, and it gives work to many people. That is something you can't say for many inventions. So, bad as it is—I don't know—it is the best invention, I think, that has ever been.[3]

We do not need to swallow Rogers's endorsement of radio whole to appreciate the line he walked between recognizing the power of a mass system and respecting the individual's limited autonomy within that system. It was one of the recurring dilemmas of twentieth-century America. In 1930, Jane Ziegler of Saugerties, New Jersey, thanked news commentator H. V. Kaltenborn for providing news briefs that made it possible for busy people to stay abreast of current events.[4] We might—and some in the 1930s essentially did—condemn the simplified and predigested version of the news Ziegler received, a standardized version mass-produced by a central authority. And we would be right to do so. But we could also reasonably join Ziegler and applaud radio for giving her access to news at all, access that may have at times enabled her as an individual to make more informed decisions about her own life. Certainly, radio made it harder for Ziegler to challenge her society's dominant values, but it also made it easier for her to operate within that outlook.

The implications, of course, extend far beyond news briefs and even the 1930s. The rise of modern mass culture and individuals' efforts to find their places within it resonated through much of the century. Making radio a part of their lives in the Depression, Americans began a process of helping to shape the meanings of their mass culture. In doing so, they turned to a medium of that culture to help them seek a degree of autonomy within it. That was part of an ongoing process, one that continued throughout the era of a mass-produced, mass-consumed culture's ascendancy. Starting in the late twentieth century, though, the shape of mass culture began changing in multiple ways. Control of production narrowed further as ownership of media outlets consolidated—a point the Internet as it develops theoretically might challenge. Simultaneously, the proliferation of stations and outlets through cable television, the Internet, and other media may have enabled audiences to personalize their consumption—unless we consider the relatively uniform consumer messages the heart of what is conveyed. Even in flux, though, our culture still blurs the lines between public and private, still pushes us to consider the value of mass communication, and still opens up the meaning of democracy in contemporary life. We continue to seek self-control and ways of being heard in our world, and we continue to look to mass culture's own vehicles for help. The journalist Dorothy Thompson, writing on Orson Welles's *War of the Worlds*, was right: the story of the twentieth century—and perhaps of the twenty-first as well—is, in significant part, the story of the balance between individual authority and mass culture. And how Americans made sense of radio is the first chapter.

NOTES

Introduction

1. *War of the Worlds*, radio script, in Hadley Cantril with Hazel Gaudet and Herta Herzog, *The Invasion from Mars: A Study in the Psychology of Panic* (Princeton, NJ: Princeton University Press, 1966), 5–8.

2. *War of the Worlds*, radio script, 9–41.

3. For estimates of listeners and how many were scared, see Cantril, 47, 56–58; Robert Brown, *Manipulating the Ether: The Power of Broadcast Radio in Thirties America* (Jefferson, NC: McFarland and Company, 1998), 219. It is possible but not inevitable that with a greater audience, more people would have been frightened. Since those who panicked were often listeners not paying close attention to the program, it is also possible that had more people deliberately tuned into the broadcast—knowing up front the program they would hear—the panic would not have increased. For responses to the broadcast, see Cantril, 47–54; Brown, 219–222; Charles Jackson, "The Night the Martians Came," in *The Aspirin Age: 1919–1941*, ed. Isabel Leighton (New York: Simon and Schuster, 1949), 433–436. See also Frederick Samuels, Jr., Watervliet, New York, to Federal Communications Commission (hereafter FCC), Oct. 30, 1938; Warner Ogden, Knoxville, Tennessee, to FCC, Oct. 30, 1938; L. C. Miller, San Gabriel, California, to FCC, Oct. 31, 1938; and Paul Morton, Trenton, New Jersey, to FCC, Oct. 31, 1938; all in Box 238, General Correspondence 1927–1946, Office of the Executive Director, FCC Collection, Research Group 173, National Archives, College Park, Maryland (hereafter "General Correspondence, FCC").

4. Dorothy Thompson, "On the Record: Mr. Welles and Mass Delusion," *New York Herald-Tribune*, Nov. 2, 1938, 21. For an account of some of the tensions between Welles and his commercial sponsors, see Michele Hilmes,

Radio Voices: American Broadcasting, 1922–1952 (Minneapolis: University of Minnesota Press, 1997), 219–225.

5. Media scholar Jeffrey Sconce sees *War of the Worlds* in terms of radio's tendency to force "participation in a vast and possibly terrifying public sphere." Jeffrey Sconce, *Haunted Media: Electronic Presence from Telegraphy to Television* (Durham, NC: Duke University Press, 2000), 109–117. Obviously listeners in the 1930s found radio valuable as well as troubling in this respect. F. M. Moody, Venice, California, to FCC, Oct. 31, 1938, Box 238, General Correspondence, FCC. For similar listeners responses, see W. H. Halliburton, Jr., Charlotte, North Carolina, to FCC, Oct. 31, 1938; Lynn Montross, Sanbornton, New Hampshire, to FCC, Nov. 1, 1938; Joseph Moore, El Dorado, Arkansas, to FCC, Nov. 1, 1938; Martin Rooney, Farmingdale, New York, to FCC, Oct. 30, 1938; all in Box 238, General Correspondence, FCC.

6. Orson Welles quoted in Brown, 227. On stage as well, Welles often assailed fascism in various forms in the 1930s. Michael Denning, *The Cultural Front: The Laboring of American Culture in the Twentieth Century* (London: Verso, 1996), 365, 375–376.

7. Cantril, ix–xi.

8. Beaumont and Nancy Newhall, New York, New York, to FCC, Oct. 31, 1938, Box 237, General Correspondence, FCC. Many listeners shared similar feelings, but see especially Skulda Baner, Milwaukee, Wisconsin, to FCC, Oct. 31, 1938; Francis Donahue, Lansing, Michigan, to FCC, Oct. 31, 1938; and Chester Schumbert, Altoona, Pennsylvania, to FCC, Oct. 31, 1938; all in Box 237, General Correspondence, FCC.

9. There has been much written on the cultural change in the United States in the decades around the turn of the twentieth century. Around the turn of the twenty-first century, some have understood the change in terms of a revolution in communications. See, for instance, John Peter Durham, *Speaking into the Air: A History of the Idea of Communication* (Chicago: University of Chicago Press, 1999); Susan Douglas, *Listening In: Radio and the American Imagination from Amos 'n' Andy and Edward R. Murrow to Wolfman Jack and Howard Stern* (New York: Times Books, 1999); Douglas Craig, *Fireside Politics: Radio and Political Culture in the United States, 1920–1940* (Baltimore: Johns Hopkins University Press, 2000), 14–15; Sconce, *Haunted Media*, 7, 109–117; Richard Brown, *Knowledge is Power: The Diffusion of Information in Early America, 1700–1865* (New York: Oxford University Press, 1989), especially 245–296. For others on communication and these cultural changes, see Lary May, *Screening Out the Past: The Birth of Mass Culture and the Motion Picture Industry* (Chicago: University of Chicago Press, 1983); Daniel Czitrom, *Media and the American Mind: From Morse to McLuhan* (Chapel Hill: University of North Carolina Press, 1982), 3–59; Paddy Scannell, *Radio, Television, and Modern Life: A Phenomenological Approach* (Oxford: Blackwell Publishers, 1996), especially 89–91. For a more wide-ranging discussion of the turn-of-the-century cultural change see, among others, Samuel Hays, *The Response to Industrialism: 1885–1914* (Chicago: University of Chicago Press, 1957); Henry May, *The End of American Innocence: A Study of the First Years of Our Own Time, 1912–1917* (Chicago: Quadrangle Books, 1959); Robert Wiebe, *The Search for Order, 1877–1920* (New York: Hill and Wang, 1967); Warren Susman, *Culture as History: The Transformation of American Society in the Twentieth Century* (New York: Pantheon Books, 1984), especially xix–xxx, 105–121, 150–183; Alan Trachtenberg, *The Incorporation of America: Culture and Society in the Gilded Age* (New York: Hill and Wang, 1982); Lynn Dumenil, *The Modern Temper: American Culture and Society in the 1920s* (New York: Hill and Wang, 1995); Lawrence Levine,

The Unpredictable Past: Explorations in American Cultural History (New York: Oxford University Press, 1993), 191–197; Daniel Boorstin, *The Americans: The Democratic Experience* (New York: Random House, 1973); Roland Marchand, *Advertising the American Dream: Making Way for Modernity, 1920–1940* (Berkeley: University of California Press, 1985); Thomas Bender, *Community and Social Change in America* (New Brunswick, NJ: Rutgers University Press, 1978), 61–137; Marshall Berman, *All That Is Solid Melts into Air: The Experience of Modernity* (New York: Simon and Schuster, 1982), 15–22; Dorothy Ross, *The Origins of American Social Science* (Cambridge: Cambridge University Press, 1991); Morton White, *Social Thought in America: The Revolt against Formalism* (Boston: Beacon Press, 1957); William Leach, *Land of Desire: Merchants, Power, and the Rise of a New American Culture* (New York: Vintage Books, 1993).

10. "For diverse social critics," historian Leo Ribuffo argues, "the slump seemed the culmination of three generations of economic, social, and moral crises." Leo Ribuffo, *The Old Christian Right: The Protestant Far Right from the Great Depression to the Cold War* (Philadelphia: Temple University Press, 1983), 3. See also Alan Brinkley, *Voices of Protest: Huey Long, Father Coughlin, and the Great Depression* (New York: Vintage Books, 1983), 144–168; Richard Pells, *Radical Visions and American Dreams: Culture and Social Thought in the Depression Years* (New York: Harper and Row, Publishers, 1973), 1–2, 25–26, 98–102, 119;

11. Lizabeth Cohen, *Making a New Deal: Industrial Workers in Chicago, 1919–1939* (Cambridge: Cambridge University Press, 1990), 100–158, 218–238, 324–333, 363–365; William Leuchtenburg, *Franklin D. Roosevelt and the New Deal, 1932–1940* (New York: Harper Torchbooks, 1963), 157–158, 331; Robert McElvaine, *The Great Depression: America, 1929–1941* (New York: Times Books, 1993), 336; Carl Degler, *Out of Our Past: The Forces That Shaped Modern America* (New York: Harper and Row, 1970), 384–393; Eric Foner, *The Story of American Freedom* (New York: W. W. Norton and Company, 1998), 196–210; John Steinbeck, The Grapes of Wrath (New York: Penguin Books, 1976), 47–50.

12. Radio, declares media historian Susan Douglas, was "arguably the most important electronic invention of the [twentieth] century." For an explanation of her rationale see Douglas, *Listening In,* 9.

13. Czitrom, 61–71; Erik Barnouw, *A History of Broadcasting in the United States,* vol. 1, *A Tower in Babel; to 1933* (New York: Oxford University Press, 1966), 9–61; Susan Douglas, *Inventing American Broadcasting, 1899–1922* (Baltimore: Johns Hopkins University Press, 1987), 142–146, 214–219, 237–249, 290–293.

14. Czitrom, 71–79; Barnouw, 61–74, 91–124, 157–160, 189–224; Douglas, *Inventing,* 300, 315–316; J. Fred MacDonald, *Don't Touch that Dial! Radio Programming in American Life from 1920 to 1960* (Chicago: Nelson-Hall, 1979), 2–25; Robert McChesney, *Telecommunications, Mass Media, and Democracy: The Battle for the Control of U.S. Broadcasting, 1928–1935* (New York: Oxford University Press, 1993), 14–27, 34; Philip Rosen, *The Modern Stentors: Radio Broadcasters and the Federal Government, 1920–1934* (Westport, CT: Greenwood Press, 1980), 9–12, 106–108, 123, 133, 140; Susan Smulyan, *Selling Radio: The Commercialization of American Broadcasting, 1920–1934* (Washington DC: Smithsonian Institution Press, 1994), 9, 38–41, 63–83.

15. McChesney, 5, 15, 29–30; Czitrom, 79–85; Rosen, 154–162, 180; Smulyan, 1–9, 94–98, 111–118, 162, 165; MacDonald, 26–40; Laurence Bergeen, *Look Now, Pay Later: The Rise of Network Broadcasting* (Garden City, NY: Doubleday and Company, 1980), 6, 8–9, 57–58; Craig, 28–35.

16. Herman Hettinger, *A Decade of Radio Advertising* (Chicago, University of Chicago Press, 1933), 42; Paul Peter, "The American Listener in 1940," *Annals of the American Academy of Political and Social Sciences* 213 (Jan. 1941), *New Horizons in Radio*, ed. Herman Hettinger, 2; Hadley Cantril with Herta Herzog and Hazel Gaudet, *The Invasion from Mars: A Study in the Psychology of Panic* (Princeton, NJ: Princeton University Press, 1940), xii. Historians Daniel Cztirom and Alice Marquis present the figures cited as an accurate portrayal of the expanding radio audience in the 1930s. Cztirom, 79; Alice Marquis, *Hope and Ashes: The Birth of Modern Times, 1929–1939* (New York: Free Press, 1996), 41. Although radio's tremendous growth during the 1930s is undisputed, not all sources agree on the exact statistics. According to the census bureau, the percentage of families with radios rose roughly from 46 percent in 1930 to 81 percent in 1940. *Historical Statistics of the United States: Colonial Times to 1970*, vol. 1–2 (Washington, DC: U.S. Bureau of the Census, 1975), 42, 796. For discussions of how much people listened to their radios, see Paul Lazarsfeld, *Radio and the Printed Page: An Introduction to the Study of Radio and Its Role in the Communication of Ideas* (New York: Duell, Sloan, and Pearce, 1940), 32; Hadley Cantril and Gordon Allport, *The Psychology of Radio* (New York: Arno Press, 1971; orig. 1935), 87–88; Herman Hettinger, "Broadcasting in the United States," *Annals of the American Academy of Political and Social Sciences* 177 (Jan. 1935), *Radio: The Fifth Estate*, ed. Herman Hettinger, 6; Peter, 3; "Radio I: A \$140,000,000 Art," *Fortune* 17, no. 5 (May 1938): 51; Frederick Lumley, *Measurement in Radio* (Columbus: Ohio State University, 1934), 196; Marquis, 41. Listeners' preferences for network programs from Cantril and Allport, 99.

17. Historian Lawrence Levine, writing about mass-produced cultural forms in general, articulates this interaction astutely. "In a modern industrial urban society," he says, "people are no more likely to be the exclusive architects of their own expressive cultures than of their own houses or clothing." But, he continues, "This is not to say that as a result people have been rendered passive, hopeless consumers. What people *can* do and *do* do is to refashion the objects created for them to fit their own values, needs, and expectations." If we want to understand culture, he rightly suggests, we must look at the interactive process between individuals and the mass-produced expressions those people receive. Levine, 293–304. There is, however, also a long and valuable line of scholars who have focused on the ways in which mass culture constrains individual choices. In the years after World War II, Frankfurt School critics often articulated this stand; more recently, students of consumer culture have usefully pointed out that consumer choice is not the same as popular consent or a choice of what to value. See, for instance, Levine, 294; Paul Gorman, *Left Intellectuals and Popular Culture in Twentieth-Century America* (Chapel Hill: University of North Carolina Press, 1996), 173–187; Leach, xiv–xv. And scholars of radio who argue that the medium's meanings exist solely in the programs or industry's structure implicitly continue this line of thought—often in very smart and subtle ways. Few histories of radio focus on the question of what radio actually meant to those who experienced the medium. There are studies of radio programs that offer valuable readings of their content, but for the most part they do not explore listeners' reactions to those programs. The level of critical analysis such studies offers varies dramatically, but see for example, Hilmes; MacDonald; and Arthur Wertheim, *Radio Comedy* (New York: Oxford University Press, 1979). Scannell offers a subtle reading of the meaning of broadcasting in British life, but, like others, he seeks to find that meaning by studying programs alone. Additionally, fine institutional histories of radio abound, but they tend to focus on the structure of the broadcasting system. See for example, Barnouw; Bergeen; Douglas, *Inventing Broadcasting*; McChesney; Rosen; and Smulyan.

The exceptions that exist are often tantalizing, but do not explore the breadth or depth of meanings Americans first found in radio. For example, Cohen and Czitrom both nicely suggest some of radio's meanings in the 1930s, but they focus their studies far more broadly; Douglas's *Listening In* likewise takes a broader sweep; moreover, she does not directly examine listeners. Craig's study likewise spends little time on listener experiences and focuses on politics exclusively, Through oral histories Ray Barfield examines how Americans saw radio, but his work betrays a nostalgia unchecked by critical analysis. Ray Barfield, *Listening to Radio, 1920–1950* (Westport, CT: Praeger Publishers, 1996). Robert Brown considers a few audience responses in his study of radio in the 1930s, but he does so to show how those on the air successfully used radio to manipulate listeners, not to get at the meanings those listeners found.

18. Thompson, 21.

Chapter 1

1. Edward Bellamy, *Looking Backward* (New York: Bantam Books, 1983), 59–62, 90–93, 151–152. The funding and programming of America's broadcasting system ultimately took a very different shape than Bellamy imagined. In his story, listeners pay for their music and lectures through a subscription fee. The novel does not explicitly indicate what determines who broadcasts, but it does suggest that anyone who can attract even a small following can publish a newspaper or preach in a church.

2. William Orton, "The Level of Thirteen-Year-Olds," *Atlantic Monthly* 147 (Jan. 1931): 3–4; see also William Orton, *America in Search of Culture* (Boston: Little, Brown, and Company: 1933), 257; James Rorty, "The Impending Radio War," *Harper's Magazine* 163 (Nov. 1931): 714, 720. Virtually all intellectuals in the early 1930s—except those involved in broadcasting—opposed the existing system in radio, historian Robert McChesney notes. Robert McChesney, *Telecommunications, Mass Media, and Democracy: The Battle for the Control of U.S. Broadcasting, 1928–1935* (New York: Oxford University Press, 1993), 86.

3. Rorty, "The Impending Radio War," 714; Rorty repeated the mirror metaphor in at least two other works in the 1930s: James Rorty, *Our Master's Voice: Advertising* (New York: The John Day Company, 1934), 267; James Rorty, *Order on the Air!* (New York: The John Day Company, 1934), 7.

4. "William Aylott Orton, 1889–1952," Faculty Meeting Minutes, Oct. 22, 1952; Smith College News Office, "Biographical Information on William A. Orton, Professor of Economics," 1950; Smith College News Office, William Orton obituary, Aug. 14, 1952; all in the Sophia Smith Collection and College Archives, Smith College, Northampton, MA; see also McChesney, 52, 87.

5. On the connections between late-nineteenth and early-twentieth-century understandings of liberalism, culture, and individualism see Herbert Hoover, *American Individualism* (Garden City, NY: Doubleday, Page and Company, 1922); Lawrence Levine, *Highbrow/Lowbrow: The Emergence of Cultural Hierarchy in America* (Cambridge, MA: Harvard University Press, 1988), 177, 200–231; Henry May, *The End of American Innocence: A Study of the First Years of Our Time, 1912–1917* (Chicago: Quadrangle Books, 1959), 30–51; Eric Foner, *The Story of American Freedom* (New York: W. W. Norton and Company, 1998), 119–123; Alan Trachtenberg, *The Incorporation of America: Culture and Society in the Gilded Age* (New York: Hill and Wang, 1982), 5, 140–163; Daniel Czitrom, *Media and the American Mind: From Morse to McLuhan* (Chapel Hill: University of North Carolina Press, 1982), 31–35. For more on intellectual journals in the 1930s, see John

Tebbel and Mary Ellen Zuckerman, *The Magazine in America, 1741–1990* (New York: Oxford University Press, 1991), 199–226.

6. Louis Filler, "Our Radio and Theirs," *Common Sense* 7, no. 8 (Aug. 1938): 12; Ring Lardner, "Heavy Da-Dee-Dough Boys," *New Yorker* 8, no. 19 (June 25, 1932): 30; Ring Lardner, "The Truth about Ruth," *New Yorker* 8, no. 20 (July 2, 1932): 27; Ring Lardner, "Lyricists Strike Paydirt," *New Yorker* 8, no. 40,(Nov. 19, 1932): 46; also see Orton, *America in Search of Culture*, 254–255; "On the Air," *Nation* 132, no. 3423 (Feb. 11, 1931): 146; Rorty, *Our Master's Voice*, 266; Otis Ferguson, "Nertz to Hertz: The Sad Air Waves," *New Republic* 78, no. 1009 (Apr. 4, 1934): 210. For more on Lardner's career as a radio critic, see Jonathan Yardley, *Ring: A Biography of Ring Lardner* (New York: Random House, 1977), 363–367.

7. Orton, *America in Search of Culture*, 246–247.

8. Orton, *America in Search of Culture*, 254, 261; William Orton, "Memorandum on Radio Policy from Professor William Orton," to the Federal Office of Education, Northampton, Massachusetts, 1934; B. H. Haggin, "The Music That Is Broadcast," *New Republic* 69, no. 894 (Jan. 20, 1932): 268; "For Better Broadcasting," *New Republic* 80, no. 1036 (Oct. 3, 1934): 201; Charles Beard and Mary Beard, *America in Midpassage*, vol. 2 (New York: Macmillan Company, 1939), 644; Merrill Denison, "Why Isn't Radio Better?" *Harper's Magazine* 168 (Apr. 1934): 582–585.

9. Orton, *America in Search of Culture*, 256–257; Orton, "The Level of Thirteen-Year-Olds," 8–9; Orton, "Memorandum on Radio Policy." For others with less fully developed, but similar, versions of this argument, see Filler, "Our Radio and Theirs," 12; Ruth Brindze, *Not to Be Broadcast: The Truth about Radio* (New York: DaCapo Press, 1974; reprint of Vanguard Press, 1937), 91; Jerome Davis, *Capitalism and Its Culture* (New York: Farrar and Rinehart 1935), 318.

10. Orton, "The Level of Thirteen-Year-Olds," 8–10; Orton, "Memorandum on Radio Policy."

11. Orton, "The Level of Thirteen-Year-Olds," 7.

12. Orton, "The Level of Thirteen-Year-Olds," 4; Orton, "Memorandum on Radio Policy"; Davis, 330.

13. Beard and Beard, 647–649; Orton, "Memorandum on Radio Policy"; see also Ring Lardner, "The Perfect Radio Program," *New Yorker* 9, no. 27 (Aug. 26, 1933): 31. For a discussion of the Victorian sense of separate cultural and economic spheres, and the Victorian idea of culture in general, see again Levine, 177, 200–231; May, 30–51; Trachtenberg, 5, 140–163.

14. Orton, *America in Search of Culture*, 264; Rorty, "The Impending Radio War," 720; Orton, "The Level of Thirteen-Year-Olds," 4, 7–9; see also Davis, 318.

15. Orton, "The Level of Thirteen-Year-Olds," 6; Orton, *America in Search of Culture*, 255–256; Travis Hoke, "Radio Goes Educational," *Harper's Magazine* 165 (Sept. 1932): 471–474; Neil Postman, *Amusing Ourselves to Death: Public Discourse in the Age of Show Business* (New York: Penguin Books, 1985), 142–154; see also "Conclusions," *New Republic* 90, no. 1163 (Mar. 17, 1937): 189.

16. Hoke, 474; for a useful account of the battle in the early 1930s over creating a radio system that would benefit education, see McChesney, whose book recounts the struggle in detail.

17. Orton, "The Level of Thirteen-Year-Olds," 5–7; Orton, *America in Search of Culture*, 255–256; "The Radio and Education," *New Republic* 63, no. 819 (Aug. 13, 1930):

357; unlike many of radio's critics, though, *New Republic* believed that radio might have a role to play in education—if the system were dramatically reformed. For other critiques of passive learning by radio, see Hoke, 474; Herbert Simpson, "Asses' Ears," *Atlantic Monthly* 148 (Dec. 1931): 777.

18. William Orton, "Radio and the Public Interest," *Atlantic Monthly* 157 (Mar. 1936): 351; William Orton, "Culture and Laissez-Faire," *Atlantic Monthly* 155 (June 1935): 752. Obviously, other developing mediums—phonographs for instance—contributed to this evolution in leisure. But none were nearly as widespread as broadcasting would be, starting in the 1930s. See Daniel Boorstin, *The Americans: The Democratic Experience* (New York: Vintage Books, 1974), 382–386.

19. Bellamy, 59–62.

20. Orton, *America in Search of Culture*, 13–15. For echoes of Orton's liberalism elsewhere in the eras, see Hoover, 14, 21–27. For a discussion of the increasing standardization and unification Orton described, see Robert Wiebe, *The Search for Order, 1877–1920* (New York: Hill and Wang, 1967), especially xiii–xiv.

21. Orton, "The Level of Thirteen-Year-Olds," 7; Orton, *America in Search of Culture*, 257; Simpson, 779; Irving Fineman, "What Are We Coming To?" *Harper's Magazine* 171 (Aug. 1935): 379; for the same critique from the political left, see Davis, 331–332; "Week-end in the Country," *New Republic* 100, no. 1293 (Sept. 13, 1939): 144.

22. Republished several times with minor tweaking, Dwight Macdonald's influential essay "Masscult and Midcult" articulated many of Orton's concerns about mass culture for readers in Post–World War II America. Dwight Macdonald, *Against the American Grain* (New York: Random House, 1962), 3–75, particularly 8–13, 36–40; Paul Gorman, *Left Intellectuals and Popular Culture in Twentieth-Century America* (Chapel Hill: University of North Carolina Press, 1996), 167–173. For examples of other cold war thinkers writing broadly on these issues of American conformity, see Louis Hartz, *The Liberal Tradition in America: An Interpretation of American Political Thought since the Revolution* (New York: Harcourt, Brace and World, 1955), 10–12, 31, 139–142, 225–226; William Whyte, Jr., *The Organization Man* (New York: Simon and Schuster, 1956), 3–14, 280–298, 393–398; David Riesman with Nathan Glazer and Reuel Denney, *The Lonely Crowd: A Study of the Changing American Character*, abridged edition (Garden City, NY: Doubleday and Company, 1953), 34–42.

23. For discussions of leftist critiques of popular culture in general in the 1930s, see Gorman, 108–110, 120–121, 138–143.

24. For more on the mainstream Left, see Richard Pells, *Radical Visions and American Dreams: Culture and Social Thought in the Depression Years* (New York: Harper and Row, Publishers, 1973), 49–50, 94–95, 395–397; for more on left-wing journals, see Tebbel and Zuckerman, 203–209.

25. Introduction to the James Rorty Collection guide, Oct. 1974, the James Rorty Collection, University of Oregon Library, Eugene, Oregon; Christopher Phelps, *Young Sidney Hook: Marxist and Pragmatist* (Ithaca, NY: Cornell University Press, 1997), 90–91, 147; McChesney, 63, 87.

26. Rorty, *Order on the Air!* 10; Brindze, *Not to Be Broadcast*, 4; Davis, 315–316.

27. Rorty, *Our Master's Voice*, 14; Brindze, *Not to Be Broadcast*, 15; Rorty, *Order on the Air!* 8–9; James Rorty, "Free Air," *Nation* 134, no. 3479 (Mar. 9, 1932): 280–281; see also "On the Air," 146; Heywood Broun, "Sponsorship," *Nation* 141, no. 3660 (Aug. 28, 1935): 190.

28. Ruth Brindze, "Who Owns the Air?" *Nation* 144, no. 17 (Apr. 17, 1937): 430; Brindze, *Not to Be Broadcast*, 11, 26–28; Rorty, *Order on the Air!* 24; Rion Bercovici, "Who Owns the Air?" *Common Sense* 2, no. 1 (July 1933): 23–24.

29. George Kaufman, "God Gets an Idea," *Nation* 146, no. 8 (Feb. 19, 1938): 208; see also Davis, 321, 327; Brindze, *Not to Be Broadcast*, 263–266; Rorty, *Order on the Air!* 22; Orton, "Radio for Robots," 198; Heywood Broun, "Labor Static," *Nation* 141, no. 3658 (Aug. 14, 1935): 190.

30. "The Week," *New Republic* 66, no. 848 (Mar. 4, 1931): 58; Rorty, "Free Air," 281; see also Rorty, "The Impending Radio War," 715; Brindze, *Not to Be Broadcast*, 90, 287–288.

31. Rorty, *Our Master's Voice*, 16; Rorty, *Order on the Air!* 25; Orton, "Radio for Robots," 195; Bercovici, 25; Brindze, *Not to Be Broadcast*, 97; T. R. Carskadon, "Radio: A Progress Report," *New Republic* 80, no. 1031 (Aug. 29, 1934): 71; William Orton, "Unscrambling the Ether," *Atlantic Monthly* 147 (Apr. 1931): 438.

32. Rorty, *Order on the Air!* 12; Brindze, *Not to Be Broadcast*, 181.

33. Brindze, *Not to Be Broadcast*, 7, 287–288; Rorty, *Our Master's Voice*, 17–18; see also Davis, 316.

34. Beard and Beard, 650; "For Better Broadcasting," 202; Filler, 11; Brindze, *Not to Be Broadcast*, 7–8, 3–4, 251–253, 258, 287–289. For a clear example of this economic understanding of fascism, see Raymond Gram Swing, *Forerunners of American Fascism* (Freeport, NY: Books for Libraries Press, 1969; orig. 1935), 22.

35. Rorty, "The Impending Radio War," 720; Rorty, *Order on the Air!* 14–15; Davis, 334. Mass-consumption critics such as William Orton joined the leftist commentators in this stand. See Orton, "Radio for Robots," 198; Orton, "The Level of Thirteen-Year-Olds," 4.

36. Bruce Bliven, "An English Miscellany," *New Republic*, 79, no. 1027 (Aug. 8, 1934): 342; Davis 330; Rorty, "Free Air," 280. When making his official pitches to reform radio, even Orton adopted a similar position. In his memo on radio for the Federal Office on Education, Orton suggested that if the federal government controlled broadcasting, it could serve an educational purpose. To turn the air into an educational medium, he wrote, the United States needed a nonprofit system that made no effort to appeal to the mass of listeners, not unlike the BBC. Orton, "Memorandum on Radio Policy."

37. Rorty, "The Impending Radio War," 722–723.

38. Rorty, "The Impending Radio War," 715.

39. Rorty, "The Impending Radio War," 714; Rorty, *Order on the Air!* 7. This concept of cultural lag was common among intellectuals in the 1930s. See John Dewey, "Toward a New Individualism," *New Republic* 62, no. 794 (Feb. 19, 1930): 13–16; Dorothy Ross, *The Origins of American Social Science* (Cambridge: Cambridge University Press), 443–444; Pells, 119.

40. Moreover, most of the articles in the *Crisis* were success stories about African Americans who flourished to some degree on radio (which, given how unwilling the networks were to give airtime to blacks, explains the lack of articles), not evaluations of the medium itself.

41. William Walls in "Bishop Flays Race Citizens Who Laugh at Amos 'n' Andy," *Pittsburgh Courier*, May 3, 1930, sec. 2, p. 1; George Schuyler, "Views and Reviews," *Pittsburgh Courier*, May 24, 1930, sec. 2, p. 12. Schuyler and the paper's publisher,

Robert Vann, were the *Courier's* most well-known and nationally respected commentators in the early 1930s. Roland Wolseley, *The Black Press, U.S.A.* (Ames: Iowa State University Press, 1971), 49.

42. Floyd Calvin, "Funny Side of Radio Broadcast 'Clowns' Is Belittled by Writer," *Pittsburgh Courier*, June 6, 1931, 1; "Amos 'n' Andy," *Pittsburgh Courier*, Apr. 25, 1931, 10; "Will Go to Radio Commission with Amos 'n' Andy," *Pittsburgh Courier*, May 9, 1931, 1; Holloway, "After Two Years of Reckless Driving," cartoon with blurb, *Pittsburgh Courier*, May 16, 1931, sec. 2, p. 2. William Walls made a similar point in a statement the paper reprinted; see William Walls, "Handicap of Amos 'n' Andy Is Related," *Pittsburgh Courier*, May 23, 1931, 1, 4.

43. See, for instance, "3,000 Preachers to Talk on Self Respect," *Pittsburgh Courier*, Oct. 24, 1931, 1.

44. George Schuyler, "Views and Reviews," *Pittsburgh Courier*, Aug. 30, 1930, 10; Holloway.

45. Ivan Earle Taylor, "The Negro Arrives," *The Crisis: A Record of the Darker Races* 40, no. 11 (Nov. 1933): 249–250; John Silvera, "Still in Blackface," *The Crisis: A Record of the Darker Races* 46, no. 3 (Mar. 1939): 77.

46. Langston Hughes to Erik Barnouw, Mar. 27, 1945, Erik Barnouw Folder, Box 10, General Correspondence, Langston Hughes Papers, James Weldon Johnson Collection in the Yale Collection of American Literature, Beinecke Rare Book and Manuscript Library, New Haven, Connecticut; Davidson Taylor to Maxim Lieber, Mar. 22, 1940, Box 43, General Correspondence, Langston Hughes Papers. Additionally, in 1936 and 1937 Hughes was recruited to write a play for a possible CBS series on America's resources; Hughes's proposed drama on black workers never aired. The early negotiations appear in Lucile Charles to Langston Hughes, Oct. 8, 1936; Langston Hughes to Lucile Charles, Feb. 18, 1937; Lucile Charles to Langston Hughes, Mar. 18, 1937; all in Box 43, General Correspondence, Langston Hughes Papers.

47. Orton, "Unscrambling the Ether," 429; Rorty, *Order on the Air!* 7, 10. Radio-minded intellectuals were not alone making such an observation. John Dewey, for instance, voiced a similar point. Pells, 119.

48. Orton, *America in Search of Culture*, 6, 246–251; Orton, "The Level of Thirteen-Year-Olds," 9–10; Orton, "Memorandum on Radio Policy"; Rorty, *Order on the Air!* 10–11. See also Bercovici, 25. In Orton's case, his own evolving thinking represented the evolution of liberalism in the era: very roughly, from the idea of a restrained government to the idea of a government actively planning for and intervening in society.

49. Orton, "Memorandum on Radio Policy"; "Can Radio Be Rescued?" *New Republic* 67 (June 24, 1931): 140; "On the Air," 146; Bercovici, 25; Bliven, 342; Filler, 10–13.

50. Rorty, *Order on the Air!* 10–11, 29–30; Brindze, *Not to Be Broadcast*, 297–305. See also Broun, "Shoot the Works," *New Republic* 94, no. 1213 (Mar. 2, 1938): 98; "The Week," *New Republic* 74, no. 954 (Mar. 15, 1933): 115; "The Week," *New Republic* 91, no. 1172 (May 19, 1937): 30; "Radio Does Its Bit," *New Republic* 100, no. 1292 (Sept. 6, 1939): 114; Orton, "Radio and the Public Interest," 356. Clearly the idea of fostering both public and private enterprise borrowed from the thinking of some New Deal programs. Given that inspiration, it is worth noting that advocates of such a plan, such as Heywood Broun, revealed a lack of faith in government as they voiced their schemes: these thinkers periodically expressed concern that government might not run radio benevolently if given full control of the air as in a BBC-style system.

It is also worth noting that *New Republic* editors particularly voiced their displeasure with radio's news reporting during the coverage of Franklin Roosevelt's first inauguration, the explosion of the Hindenburg, and the opening of war in Europe. They were frustrated that news only came in irregularly timed flashes. Ironically, coverage of the latter two events has been credited with helping to win radio acclaim as a news source. Robert Brown, *Manipulating the Ether: The Power of Broadcast Radio in Thirties America* (Jefferson, NC: McFarland and Company, 1998), 141–144, 175–180. The solution *New Republic* called for—stations arrayed by content, one for news, one for serious music, and so on—was not explicitly a public-private competition but it also amounted to noncommercial diversification of the air.

51. Brindze, *Not to Be Broadcast*, 297; Bliven, 342. Bliven was not entirely sanguine about the power of the state. He simply felt that the federal government already had the power to keep radicals in check; explicit control of radio would not change that.

52. McChesney, 86.

53. Leopold Stokowski, "New Vistas in Radio," *Atlantic Monthly* 155 (Jan. 1935): 8, 16.

54. Deems Taylor, *The Well Tempered Listener* (New York: Simon and Schuster, 1940), 269–270; Gilbert Seldes, *Mainland* (New York: Charles Scribner's Sons, 1936), 34; see also John Tasker Howard, "Better Days for Music," *Harper's Magazine* 174 (Apr. 1937): 485; Jascha Heifetz, "Radio, American Style," *Harper's Magazine* 175 (Oct. 1937): 500–501; Dickson Skinner, "Music Goes into Mass Production," *Harper's Magazine* 178 (Apr. 1939): 487.

55. Seldes, 36; Deems Taylor, "Radio—A Brief for the Defense," Harper's Magazine 166 (Apr. 1933): 561; Richard Sheridan Ames, "The Art of Pleasing Everybody," *Atlantic Monthly* 158 (Oct. 1936): 445, 448, 451; Merrill Denison, "Freedom, Radio, and the FCC," *Harper's Magazine* 178 (May 1939): 635; Heifetz, 497; Taylor, *The Well Tempered Listener*, 266–268.

56. Denison, "Freedom, Radio, and the FCC," 635; Seldes, 35–38.

57. William Hard, "Europe's Air and Ours," *Atlantic Monthly* 150 (Oct. 1932): 509. See also Taylor, "Radio—A Brief for the Defense," 559–560.

58. Denison, "Why Isn't Radio Better?" 585–586; Denison, "Freedom, Radio, and the FCC," 635–636, 640; Merrill Denison, "Soap Opera," *Harper's Magazine* 180 (Apr. 1940): 505.

59. T. R. Carskadon, "Radio: The Happy Slattern," *New Republic* 92, no. 1190 (Sept. 22, 1937): 183–184; Untitled, *Nation* 147, no. 14 (Oct. 1, 1938): 310; James Rorty, "Radio Comes Through," *Nation* 147, no. 16 (Oct. 15, 1938): 372–374. Even in the United States, Rorty said on the eve of World War II, the greatest danger of censorship came from the U.S. government: in the event of a national crisis, he worried, the president would take control of the airwaves for his own purposes.

Chapter 2

1. Janet Bonthuis to *Vox Pop*, Nov. 20, 1939; Janet Bonthuis to *Vox Pop*, no date; Janet Bonthuis to *Vox Pop*, Dec. 29, 1939; all in Folder 7, Box 1, Series I, Vox Pop Collection, Broadcast Pioneers Library of American Broadcasting, University of Maryland, College Park, MD (hereafter Vox Pop Collection).

2. Radio was not alone in fostering socially constructed communities in the twentieth century; such communities, once rare, became familiar as the century progressed.

Indeed, in the Depression many turned to such wide-reaching organizations—the CIO in Lizabeth Cohen's study, for instance—for the support local communities could no longer provide. But radio was the most widespread source—including enormous numbers of participants—and perhaps the most distinctively modern: the bonds uniting radio communities were literally invisible and almost never manifested themselves through actual face-to-face encounters. Moreover, radio enabled Americans to build such communities in a leisure activity, not just in the work-based or political organizations many historians explore. For discussions of the rise of nongeographic communities early in the twentieth century see Lynn Dumenil, *The Modern Temper: American Culture and Society in the 1920s* (New York: Hill and Wang, 1995), 42; Daniel Boorstin, *The Americans: The Democratic Experience* (New York: Random House, 1973), 89–90, 119, 122–123, 147; Thomas Bender, *Community and Social Change in America* (New Brunswick, NJ: Rutgers University Press, 1978), 108–111; Lizabeth Cohen, *Making a New Deal: Industrial Workers in Chicago, 1919–1939* (Cambridge: Cambridge University Press, 1990), 249–253, 292–293.

3. For RCA radio prices, see "The New RCA Victor Line," 1931; and "RCA Victor Radios for 1939," Folders 14 and 21, respectively, in Box 10, Series VII, RCA Papers, Accession #2069, Hagley Museum and Library, Wilmington, DE (hereafter RCA Papers). For listener estimates, see Herman Hettinger, *A Decade of Radio Advertising* (Chicago: University of Chicago Press, 1933), 42; Paul Peter, "The American Listener in 1940," *Annals of the American Academy of Political and Social Sciences* 213 (Jan. 1941), *New Horizons in Radio*, ed. Herman Hettinger, 2; Hadley Cantril with Herta Herzog and Hazel Gaudet, *The Invasion from Mars: A Study in the Psychology of Panic* (Princeton, NJ: Princeton University Press, 1940), xii; Daniel Czitrom, *Media and the American Mind: From Morse to McLuhan* (Chapel Hill: University of North Carolina Press, 1982), 79; Alice Marquis, *Hope and Ashes: The Birth of Modern Times 1929–1939* (New York: Free Press, 1996), 41. The U.S. Census Bureau's figures show the same trend a bit less dramatically, finding that the percentage of families with radios rose roughly from 46 percent in 1930 to 81 percent in 1940. *Historical Statistics of the United States: Colonial Times to 1970*, vol. 1–2 (Washington, DC: U.S. Bureau of the Census, 1975), 42, 796.

4. "Favorite Recreations," *Fortune* 17, no. 1 (Jan. 1938): 88; "The Movies," *Fortune* 20, no. 5 (Nov. 1939): 176; Erik Barnouw, *A History of Broadcasting in the United States*, vol. 2, *The Golden Web: 1933–1953* (New York: Oxford University Press, 1968), 6; unemployed man quoted in Frances Holter, "Radio among the Unemployed," *Journal of Applied Psychology* 22, no. 1 (Feb. 1939): 164. Studies among children also found radio very popular. Azriel Eisenberg, *Children and Radio Programs: A Study of More Than Three Thousand Children in the New York Metropolitan Area* (New York: Columbia University Press, 1936), 80–81, 184; Sidonie Gruenberg, *Radio and Children* (New York: Radio Institute of the Audible Arts, 1935), 4.

5. Edward Jacobsen to John Royal, Sept. 14, 1934; P. G. Parker to Alfred Morton and Edgar Kobak, interdepartmental memo, Sept. 26, 1934, both in Folder 5, Box 24, NBC Papers, State Historical Society of Wisconsin, Madison, WI. Commissioner Kenesaw Mountain Landis rejected Tyson on the grounds that the Detroit announcer might be biased for his hometown team. The 1934 World Series also claimed radio fame as the first World Series to be presented as a sponsored broadcast: Ford Motor Company paid for the games. Prior to 1934, the network presented the series without advertisements, as though carrying the baseball championship enabled radio to meet its legal obligation to provide broadcasting that served the public.

6. Ruth Brindze, *Not to Be Broadcast: The Truth about the Radio* (New York: DaCapo Press, 1974; orig. 1937), 26–28; Hadley Cantril and Gordon Allport, *The Psychology of Radio* (New York: Arno Press, 1971; orig. 1935), 99. Many historians looking at radio have emphasized the medium as a vehicle homogenizing the country. See for example Marquis, 7, 47; Alice Marquis, "Written on the Wind: The Impact of Radio during the 1930s," *Journal of Contemporary History* 19, no. 3 (July 1984): 411; Michele Hilmes, *Radio Voices: American Broadcasting, 1922–1952* (Minneapolis: University of Minnesota Press, 1997), 5; Susan Smulyan, *Selling Radio: The Commercialization of American Broadcasting, 1920–1934* (Washington, DC: Smithsonian Institution Press, 1994), 63–64, 162, 165. Lizabeth Cohen makes a similar assertion, but she rightly qualifies it by noting that the emerging mass culture did not include all classes evenly. Cohen, 325–331.

7. Paul Lazarsfeld, *Radio and the Printed Page: An Introduction to the Study of Radio and Its Role in the Communication of Ideas* (New York: Duell, Sloan, and Pearce, 1940), 15–35; H. M. Beville, Jr., *Social Stratification of the Radio Audience* (A Study Made for the Princeton Radio Research Project, 1939), 18–23, 34–42; Cantril and Allport, 87; Herman Hettinger and Walter Neff, *Practical Radio Advertising* (New York: Prentice-Hall, 1938), 111–115. Historians Lizabeth Cohen and J. Fred MacDonald both note the class elements in radio's appeal. According to Cohen, the emerging mass culture of the twentieth century was, in fact, a class culture: even as it won a widespread audience, radio maintained a particular class appeal, she argues. As MacDonald sees it, the effort to attract as large an audience as possible meant radio courted a broadly defined group of middle-class listeners, indicating for him that it was a democratic medium in which the majority ruled. His conclusion ignores the question of who actually selected the programs for that mass audience. Cohen, 329–330; J. Fred MacDonald, *Don't Touch that Dial! Radio Programming in American Life, 1920 to 1960* (Chicago: Nelson-Hall, 1979), 2, 90.

8. Cantril and Allport, 85–86; Ida Moore, "A Little Amusemint," *These Are Our Lives: As Told by the People and Written by Members of the Federal Writers' Project of the Works Progress Administration in North Carolina, Tenne., and Georgia* (New York: W. W. Norton and Company, 1939), 221. Only about one in seven of the individuals in the collection said they had radios. Edmund deS. Brunner, *Radio and the Farmer and a Symposium on the Relation of Radio to Rural Life* (New York: Radio Institute of the Audible Arts, 1935), 17–18, 29; Hettinger, *A Decade of Radio Advertising*, 48–49. According to a 1938 *Fortune* magazine survey, roughly half of African American homes had radios, compared to 88 percent of the population in general. The *Fortune* study's estimates of radio's distribution exceeded others at the time, but the magazine's point about the significant racial disparity was valid. "Toscanini on the Air" *Fortune* 17, no. 1 (Jan. 1938): 62. For late 1930s geographic distribution of sets, see Peter, 2; Lazarsfeld, 267. They noted that at the end of the decade, when over 90 percent of urban homes had radios, less than 70 percent of rural homes—and just over 60 percent of all southern homes—did.

9. Georgian farmer quoted in Charles Wolfe, "The Triumph of the Hills: Country Radio, 1920–1950," In *Country: The Music and the Musicians*, ed. Paul Kingsbury, Alan Axelrod, and Susan Costello (New York: Abbeville Press, 1994), 42. Memories of African Americans in the South coming together to listen to Joe Louis fight on the radio recur. See Clifton Taulbert, *Once Upon a Time When We Were Colored* (Tulsa, OK: Council Oak Books, 1989), 51–52; Maya Angelou, *I Know Why the Caged Bird Sings* (New York: Bantam Books, 1971), 111–115. Grapevine radio discussed in Ray Barfield, *Listening to Radio, 1920–1950* (Westport, CT: Praeger Publishers, 1996), 53–55.

10. "RCA Victor Radios for 1939," Folder 21, Box 10, RCA Papers.

11. Cohen, 133; Frederick Lumley, *Measurement in Radio* (Columbus: Ohio State University, 1934), 200; Cantril and Allport, 85; Hettinger, *A Decade of Radio Advertising*, 36. Most of the common group listening took place in families or among unorganized friends, but some formal clubs did arise to listen to specific programs as groups. According to Sherwood Gates, by 1941 there were roughly 15,000 organized listening groups with between 300,000 to 450,000 members nationwide. Barfield, 39–40; Lazarsfeld, 107; Sherwood Gates, "Radio in Relation to Recreation and Culture," *Annals of the American Academy of Political and Social Sciences* 213 (Jan. 1941), *New Horizons in Radio*, ed. Herman Hettinger, 11. By 1940 in Chicago 40 percent of working-class families had two or more radios in their home. In the decade that followed, the practice having a separate radio—or radios—for the children in a family became common. Cohen, 326; Barfield, 30.

12. As late as 1934 most New York City drivers opposed radios in cars as too dangerous; some states actually banned car radios. By the end of the decade, though, the vast majority of people surveyed approved of radio-equipped autos. Edward Suchman, "Radio Listening and Automobiles," *Journal of Applied Psychology* 23, no. 1 (Feb. 1939): 148–149. See also Columbia Broadcasting System, "An Analysis of Radio-Listening in Automobiles," Pamphlet File, Broadcast Pioneers Library of American Broadcasting (hereafter BPLAB).

13. A. D. Medoff to *Vox Pop*, Jan. 9, 1936, Folder 4, Box 1, Series I, Vox Pop Collection. Cantril and Allport, 89; Lumley, 195, 202; Herman Hettinger, "Study of Habits and Preferences of Radio Listeners in Philadelphia" (1930), 11, Folder Audience Surveys (2), Box 6, Vertical Files, BPLAB; Eisenberg, 136–137; Cantril, 80.

14. Hilmes, 154–155. For preferred types of programs by gender, see Cantril and Allport, 91–93. Additionally, sports broadcasts, which drew a largely male following, actually took up a very small portion of the airtime: never more than 2 percent through the decade. Kenneth Bartlett, "Trends in Radio Programs," *Annals of the American Academy of Political and Social Sciences* 213 (Jan. 1941), *New Horizons in Radio*, ed. Herman Hettinger, 18.

15. Columbia Broadcasting System, "An Analysis of Radio-Listening in Automobiles," 18. Most studies found that over half the airtime was devoted to musical programs, even by the end of the 1930s. Cantril and Allport, 93, 217; Bartlett, 16; FCC survey in Peter, 4. Czitrom disagrees with that figure, asserting that by 1940 music represented only a quarter of network evening programming. Czitrom, 84.

16. Sources in at least two oral histories said they remembered overhearing full *Amos 'n' Andy* programs through windows while walking: Barfield, 24; Jacquelyn Dowd Hallet al., *Like a Family: The Making of a Southern Cotton Mill World* (New York: W. W. Norton and Company, 1987), 258. At the program's height around 1930, listenership measures were very imprecise, but some estimated one-third of America heard the program nightly. Cohen, 329; MacDonald, 27; Hall, 258; Arthur Frank Wertheim, *Radio Comedy* (New York: Oxford University Press, 1979), 48–49, 55; Melvin Patrick Ely, *The Adventures of Amos 'n' Andy: A Social History of an American Phenomenon* (New York: Free Press, 1991), 5, 115–116. For a detailed discussion of the history of the program, see Ely. For a needed counterpoint about the role of race in the program, see Hilmes, 81–93.

17. MacDonald, 47–52, 78–80, 91–137, 240; Ely, 199; Hilmes, 92, 183–184; Czitrom, 86; Lazarsfeld, 204. Although quiz shows such as *Vox Pop* would become more popular after World War II, by 1937 they were a vital form of programming on the air.

18. Marquis, *Hope and Ashes*, 30; "The NBC Handbook on Offers and Contests," 1934, 1935, 1937, 1939 editions, Folder Promotions/Publicity—Merchandising Tie In, Vertical File, BPLAB; Susan Ware, *It's One O'clock and Here Is Mary Margaret McBride: A Radio Biography* (New York: New York University Press, 2005), 66. Unfortunately for us today, the letters to McBride were lost to World War II paper drives.

19. Janet Bonthuis to *Vox Pop*, Nov. 20, 1939.

20. Nate Tufts to Parks Johnson, May 14, 1940, Oct. 25, 1939, Folder Business Files and Correspondence—Ruthrauff and Ryan, Inc. (1), Series I, Vox Pop Collection; Rogers quoted in Wertheim, 63; Robert Morton in "Westinghouse Radio Stations—Using Broadcast Advertising," Folder KDKA—Using Broadcast Advertising 1930, Box 1, Station Files, BPLAB, 1–2; Frank Leroy Blanchard, "Experiences of a National Advertiser with Broadcasting," in "Westinghouse Radio Stations—Using Broadcast Advertising," 1; National Broadcasting Company, "Let's Look at Radio Together," Pamphlet File, BPLAB, 18; Cantril and Allport, 62–63, 71. Through the 1930s the majority of network airtime actually remained unsponsored. But the most popular programs were almost invariably commercial ones. In the mid-1930s when only about one-third of airtime was paid for by advertisers, radio fans tuned to sponsored programs so regularly that listeners estimated that nearly three-fourths of all programs were commercial ones.

For discussions of the commercialization of American radio, see Susan Douglas, *Inventing American Broadcasting, 1899–1922* (Baltimore: Johns Hopkins University Press, 1987), xv–xvii, 233–240, 290, 300–307, 315–321; Robert McChesney, *Telecommunications, Mass Media, and Democracy: The Battle for the Control of U.S. Broadcasting, 1928–1935* (New York: Oxford University Press, 1993); Smulyan. For a range of discussions of consumerism in the early twentieth century more generally, see, for example, Warren Susman, *Culture as History: The Transformation of American Society in the Twentieth Century* (New York: Pantheon Books, 1984), xx–xxiv, 111, 122–147; 187–188, 277–280; William Leach, *Land of Desire: Merchants, Power, and the Rise of a New American Culture* (New York: Vintage Books, 1993), xiii–xv, 3–12, 379–390; Dumenil, 56–58, 78–97; Roland Marchand, *Advertising the American Dream: Making Way for Modernity, 1920–1940* (Berkeley: University of California Press, 1985).

21. Cantril and Allport, 66, 240; Eisenberg, 63; William Robinson, "Radio Comes to the Farmer," Paul Lazarsfeld and Frank Stanton, eds., *Radio Research 1941* (New York: Duell, Sloan and Pearce, 1941), 269, 272; Joy Cole to Jane Crusinberry, Sept. 24, 1938, Folder 9, Box 3, Crusinberry Papers, State Historical Society of Wisconsin (hereafter Crusinberry Papers); Mrs. Pete Detzal to Alois Havrilla, Feb. 16, 1931, Folder Correspondence, Fan Mail, February 1931, Box 1, Alois Havrilla Collection, BPLAB; Patricia Pierson to *Vox Pop*, Feb. 19, 1938, Folder Business Files and Correspondence—Look Alike Contest, Series I, Vox Pop Collection; George Graham to *Vox Pop*, May 26, 1938, Folder Business Files and Correspondence—Facts Supporting Vox Pop Statements, Series I, Vox Pop Collection; see also Thomas Varner to *Vox Pop*, Aug. 6, 1938, Folder 6, Box 1, Series I, Vox Pop Collection. Even when interviewed decades later, some listeners acknowledged they had often purchased a company's products based on that company sponsoring a well-liked program. Barfield, 116–117.

22. Mrs. Leo Lewalski to *Vox Pop*, Feb. 16, 1937, Folder 5, Box 1, Series I, Vox Pop Collection; for other examples, see also F. C. Lussenhop to *Vox Pop*, Sept. 17, 1936, Folder 4, Box 1; H. D. Eagle to *Vox Pop*, Feb. 10, 1937, Folder 5, Box 1; Linden Scott to *Vox Pop*, Dec. 14, 1937, Folder Business Files and Correspondence—Look Alike Contest;

all in Series I, Vox Pop Collection. See also Marilyn Hoffman to Jane Crusinberry, Mar. 26, 1941; Mrs. Fred Dean to Jane Crusinberry, Mar. 27, 1941; and Anon. to Jane Crusinberry, Apr. 3, 1941, all in Folder 10, Box 3, Crusinberry Papers. Kathy Newman takes a similar argument in a more collective direction, exploring organized efforts to use listeners' consumer identities as a source of power. Kathy Newman, *Radio Active: Advertising and Consumer Activism, 1935–1947* (Berkeley: University of California Press, 2004).

23. Historians who examine commercial broadcasting from the production side, such as Smulyan and to a lesser extent McChesney, often imply, inadvertently perhaps, that radio had no meaning beyond its corporate identity. Their works are crucial to understand radio, but also reveal the problem of trying to make sense of cultural expressions purely by looking at those who produced the expressions.

24. "When You Can't Find a Friend, You've Still Got the Radio" comes from the chorus of Nanci Griffith's song "Listen to the Radio," *Storms* (MCA Records, 1989). Grace Squires to Jane Crusinberry, April 11, 1940, Folder 10, Box 3, Crusinberry Papers. For more on *The Story of Mary Marlin*, see Hilmes, 175–181.

25. We might also call these communities imagined ones, since they were largely fashioned in the minds of listeners and bound together by imagined bonds. However, the term "imagined communities" as coined by Benedict Anderson and used by several other scholars in exploring radio has a different meaning. Like Anderson's imagined communities, the ethereal communities listeners forged were built by individuals envisioning a sense of connection to and fellowship with people they never actually met. However, unlike imagined communities, which share an abstract national identity, the communities that I found listeners more commonly engaging in through the radio were concrete and personal: individual relationships with other members of the community. Listeners felt a bond with particular radio personas, not as often with the mass audience as a whole. See Benedict Anderson, *Imagined Communities: Reflections on the Origin and Spread of Nationalism* (London: Verso, 1985), particularly 15–16. Michele Hilmes offers one view on applying Anderson's thinking to radio. See Hilmes, 11–23.

26. Robert Lynd and Helen Lynd, *Middletown in Transition: A Study in Cultural Conflicts* (New York: Harcourt, Brace and Company, 1937), 491; Bender, 89, 108–119; Robert Wiebe, *The Search for Order, 1877–1920* (New York: Hill and Wang, 1967), xiii–xiv, 2–14, 22–23, 42–49, 52, 76; Robert McElvaine, *The Great Depression: America 1929–1941* (New York: Times Books, 1993), xxvi–xxix, 8–9; Boorstin, 89–90, 119, 122–123, 132–133; Dumenil, 4, 42; Lazarsfeld, 200–201; Cohen, 54–55, 95–100, 214, 249–253, 363–365; Richard Pells, *Radical Visions and American Dreams: Culture and Social Thought in the Depression Years* (New York: Harper and Row, Publishers, 1973), 108. It should be noted, as Thomas Bender does, that these pressures did not mean the destruction of community in an absolute sense; communal forms changed over time. Bender emphasizes that in the 1900s communities became both smaller and larger: Americans no longer felt connected to their entire towns, but created smaller, perhaps neighborhood-based, communities as well as translocal bonds. Wiebe, Boorstin, and Dumenil particularly discuss the idea that new, socially constructed communities began emerging in the decades around the turn of the century, taking the place of geographically based communities.

27. Robinson, 239–240; Lynd and Lynd, 264, 467; see also Brunner, 5, 11–15; Cantril and Allport, 19, 239, 259; Lazarsfeld, 201; Charles Siepmann, "Radio and Education," *Studies in Philosophy and Social Science* 9, no. 1 (1941): 106.

28. Geo. Burglehaus to H. V. Kaltenborn, June 20, 1930, Folder June-Oct. 1930, Box 17, H. V. Kaltenborn Papers, State Historical Society of Wisconsin.

29. Parks Johnson, note written on letter from Nate Tufts, July 28, 1939, Folder Business Files and Correspondence—Ruthrauff and Ryan Inc. (1), Series I, Vox Pop Collection. Sample questions the *Vox Pop* announcers asked included "Who wrote: The Gold Bug . . . King Lear . . . Ivanhoe?" "A tree is 15-feet shorter than a pole, which is three times as tall as the tree. How tall is the tree?" "Is an elevator still an elevator when it's coming down?" "Can you swallow without moving your tongue? Try it now!" and "If entertaining an out-of-town visitor in your home town, and [you] wanted to show him the town, what three things would you show him?" Folder 25, Box 1, Series I, Vox Pop Collection.

30. Writing on British broadcasting, Paddy Scannell makes the compelling argument that broadcasters portrayed the mass audience in personal terms. Clearly we can see that to be true of American radio at this time as well. But that does not mean listeners had to see the medium that way. Plenty of messages broadcasting disseminated were rejected. That broadcasters often constructed radio as an intimate communication is significant chiefly because that was how listeners chose to see the medium. Even Scannell, who does not consider listener understandings of radio, admits that broadcasters followed the lead of listener desires here. Paddy Scannell, *Radio, Television and Modern Life: A Phenomenological Approach* (Oxford: Blackwell Publishers, 1996), 2–7, 12–16, 23–24, 56. On crooning, see Terry Cooney, *Balancing Acts: American Thought and Culture in the 1930s* (New York: Twayne Publishers, 1995), 91–92; Scannell, 61–64. On Mary Margaret McBride's use of listener letters, see Ware, 66. For listener responses to premiums and radio clubs, see "The NBC Handbook on Offers and Contests" 1934, 1935, 1937, 1939 editions, particularly 1935; for WLS station marketing, see *WLS Family Album*, 1930–1937, Folder WLS Family Album, 1929–1937, Box 6, Station Files, BPLAB.

31. Parks Johnson, "General Observations." Folder 34, Box 1, Series I, Vox Pop Collection.

32. G. E. Hauson to *Vox Pop*, Feb. 12, 1938; and Harold Peters to *Vox Pop*, Dec. 18, 1937; both in Folder Listener Correspondence Geoduck, Series I, Vox Pop Collection; Ethel Strintz to *Vox Pop*, April 24, 1936, Folder Business Files and Correspondence—Facts Supporting Vox Pop Statements, Series I, Vox Pop Collection.

33. Gertrude Christine to H. V. Kaltenborn, March 1933, Folder Feb. 10 to March 20, 1933, Box 22, H. V. Kaltenborn Papers; Charles and Elizabeth Dunlap to H. V. Kaltenborn, Sept. 30, 1938, Folder Sept. 30, 1938, Box 32, H. V. Kaltenborn Papers; Ware, 65; Wertheim, 89; Wolfe, 48.

34. Herta Herzog, "On Borrowed Experience: An Analysis of Listening to Daytime Sketches," *Studies in Philosophy and Social Science* 9, no. 1 (1941): 67, 85. Jessie Penn to Jane Crusinberry, no date, Folder 9; Beulah deRocher to Jane Crusinberry, April 7, 1937, Folder 10; Howard Matteson to Jane Crusinberry, Oct. 8, 1937, Folder 10; and Alice Neale to Jane Crusinberry, April 10, 1941, Folder 10, all in Box 3, Crusinberry Papers.

35. Catherine White to Jane Crusinberry, May 14, 1941, Folder 9, Box 3, Crusinberry Papers; Mrs. Lefkowitz interview answers, B0130, Folder Questionnaires, Bureau of Applied Social Research Collection, Lehman Library, Columbia University, New York; Grace Roggenstein to Jane Crusinberry, no date, Folder 9, Box 3, Crusinberry Papers; Herzog, "On Borrowed Experience," 68, 83.

36. "'Amos 'n' Andy' Aid in Taxi Man's Freedom," *Pittsburgh Courier*, April 26, 1930, sec. 2, p. 1; Ely, 144–147; Marquis, *Hope and Ashes*, 29. To many white listeners of *Amos*

'*n'Andy*, while radio made the supposedly African American world of the program seem real and personal, the negative radio portrayal of that world tamed it, and made it unreal enough that many whites were not made uncomfortable empathizing with it. Even so, there is nothing in the scant listeners' responses to *Amos 'n'Andy* I have seen to suggest white listeners considered their relationships with the characters to be relationships among equals. It is also possible that the fact that the actors playing Amos and Andy and the other characters were whites may have influenced some listeners' responses to the program. Clearly, the popularity of *Amos 'n'Andy* raises all sorts of questions of race in America, questions beyond the scope of my sources and this project. See Hilmes, 81–96.

37. Cantril and Allport, 234; Herta Herzog, "Survey of Research on Children's Radio Listening," MS, Office of Radio Research, Columbia University, New York, 39–40, 76.

38. George Roy Brown to *Vox Pop*, Aug. 3, 1938, Folder 6; and Gordon Bockus to *Vox Pop*, May 3, 1940, Folder 8; both in Box 1, Series I, Vox Pop Collection. Mrs. Feldman to H. V. Kaltenborn, April 29, 1930, Folder April to May 1930, Box 17; Louise Newhall to H. V. Kaltenborn, Feb. 10, 1933, Folder Feb. 10 to March 20, 1933, Box 22; and Concha Marin to H. V. Kaltenborn, Sept. 30, 1938, Folder Sept. 30, 1938, Box 32; all in H. V. Kaltenborn Papers; Cantril and Allport, 109–126.

39. Brooklyn woman and Manhattan mom in Eisenberg, 154–155; Cantril and Allport, 102; Ray to *Vox Pop*, Nov. 27, 1937, Folder Business Files and Correspondence— Look Alike Contest, 1938, Series I, Vox Pop Collection; Harold Johnson in Marchand, 93.

40. Lily Sykes to H. V. Kaltenborn, April 29, 1930, Folder April to May 1930, Box 17, H. V. Kaltenborn Papers. Everett Snelling to *Vox Pop*, April 20, 1937, Folder 5; for replies to Snelling's claim, see Folder 7; Pinto Colvig to *Vox Pop*, Oct. 22, 1937, Folder 5; all in Box 1, Series I, Vox Pop Collection; Herzog, "On Borrowed Experience," 86.

41. William Vogel to *Vox Pop*, Dec. 21, 1936, Folder 4; Emily Pierce to *Vox Pop*, Oct. 20, 1939, Folder 7; Pinto Colvig to *Vox Pop*, Oct. 22, 1937, Folder 5; all in Box 1, Series I, Vox Pop Collection. In 1933 Colvig had written the lyrics to the Disney musical film *The Three Little Pigs*. In his song "Who's Afraid of the Big Bad Wolf," Colvig had to repeat the same line several times because he could not come up with a word to rhyme with "wolf." The block apparently still haunted him four years later when he turned to *Vox Pop* seeking a rhyme and relief. Announcer Parks Johnson noted this same tendency of listeners to write to the program with any questions that had stumped them. For instance, a teenage girl wrote to him asking for an interpretation of one of her dreams, he recalled. Johnson, "General Observations," Folder 34, Box 1, Series I, Vox Pop Collection.

42. Betty Chalk to H. V. Kaltenborn, Feb. 23, 1933, Folder Feb. 10 to March 20, 1933, Box 22; George Hull to H. V. Kaltenborn, Feb. 23, 1933, Folder Feb. 10 to March 20, 1933, Box 22; and Inez Pelins to H. V. Kaltenborn, June 23, 1930, Folder June to Oct. 1930, Box 17; all in H. V. Kaltenborn Papers; *WLS Family Album*, 1930, Folder WLS Family Album, 1929–1937, Box 6, Station Files, 15.

43. Esther Norman to Jane Crusinberry, April 5, 1940; and Catherine White to Jane Crusinberry, May 14, 1941; both in Folder 9, Box 3, Crusinberry Papers; Mrs. Ryan interview answers, B0130, Folder Questionnaires, Bureau of Applied Social Research Collection; Herzog, "On Borrowed Experience," 74, 78, 86, 89–90.

44. Herzog, *Children's Radio Listening*, 21, 57, 65–66, 80; Gruenberg, 5; Doris Hough to H. V. Kaltenborn, June 12, 1930, Folder June to Oct. 1930, Box 17, H. V. Kaltenborn Papers. There was actually a popular program in the 1930s devoted to this project: giving listeners advice on interpersonal matters. *The Voice of Experience* received ten to twenty

thousand letters from listeners each week. Most listeners received one of approximately a hundred standardized printed responses. *The Voice* discussed the more dramatic problems on the air. Erik Barnouw, *A Tower in Babel: A History of Broadcasting*, vol. 1, *A Tower in Babel; to 1933*, (New York: Oxford University Press, 1966), 241.

45. C. P. Lohse to H. V. Kaltenborn, Feb. 13, 1933, Folder Feb. 10 to March 20, 1933, Box 22, H. V. Kaltenborn Papers.

46. Cohen, 325–331; Cantril and Allport, 18; Holter, 164. As Cohen points out, these were new kinds of communities in some respects, particularly in terms of who partici-pated in them. Cohen explains that the common culture of radio could help integrate groups of people who might not have interacted in the past. However, the shape of these communities could remain familiar: many were still personal networks connected, at least some of the time, by face-to-face relations. Michele Hilmes observes a similar phenom-enon regarding radio, though on a broader scale, and, of course, these are the sorts of imagined bonds Benedict Anderson refers to in his discussion of new kinds of national communities. See Hilmes, 11–13; and Anderson.

47. Eisenberg 162; Herzog, *Children's Radio Listening*, 33; Cohen 133; Bill Oliver to *Vox Pop*, Dec. 6, 1937, Folder 5, Box 1, Series I, Vox Pop Collection; Julia Bingham to H. V. Kaltenborn, Feb. 12, 1933, Folder Feb. 10 to March 20, 1933, Box 22, H. V. Kalten-born Papers; Holter, 166.

48. Andrew McQuaker to *Vox Pop*, Dec. 7, 1937, Folder 5, Box 1, Series I, Vox Pop Collection. Ralph Graves to H. V. Kaltenborn, Sept. 23, 1938, Folder Sept. 29, 1938; C. J. Sweeney to H. V. Kaltenborn, Sept. 29, 1938, Folder Sept. 29, 1938; both in Box 32, H. V. Kaltenborn Papers. Eisenberg 52, 163, 194; Herzog, *Children's Radio Listening*, 66, 73–74. According to one study, radio was at times the only way for poor urban families to continue to participate in the culture of their neighbors and associates. Holter, 164.

49. Cantril and Allport, 102; Robinson, 274–277; Herzog, *Children's Radio Listening*, 53.

50. Agnes Thompson to H. V. Kaltenborn, Sept. 30, 1938, Folder Sept. 30, 1938, Box 32, H. V. Kaltenborn Papers.

51. For more on relevant ideas of the public sphere, see Michael Warner, *Letters of the Republic: Publication and the Public Sphere in Eighteenth Century America* (Cambridge, MA: Harvard University Press, 1990), 34–69; Mary Ryan, *Civic Wars: Democracy and Public Life in the American City during the Nineteenth Century* (Berkeley: University of California Press, 1997), 5–13; John Durham Peters, *Speaking into the Air: A History of the Idea of Communication* (Chicago: University of Chicago Press, 1999), 210–211; and, of course, Jürgen Habermas, *The Structural Transformation of the Public Sphere: An Inquiry into a Category of Bourgeois Society*, trans. Thomas Burger (Cambridge, MA: MIT Press, 1989), xi–xii, 6–26. Much of this discussion of the public sphere considers the idea before the advent of radio, but it nonetheless offers a useful frame to see the arrival of a very different means of creating that sphere.

52. Grace Squires to Jane Crusinberry, April 11, 1940, Folder 10, Box 3, Crusinberry Papers.

Chapter 3

1. William Nels, Pursglove, West Virginia, to Franklin Roosevelt, April 28, 1935, Box 21, PPF 200, Franklin D. Roosevelt Library, Hyde Park, New York (hereafter FDR Library). Nels's experience doubly reveals the importance of Roosevelt's fireside chat in

bridging the distance that otherwise separated listeners from the federal government: Nels actually obtained work through a New Deal relief agency, but he still felt no meaningful connection to the national government before listening to Roosevelt's radio address.

2. Michael McGerr, *The Decline of Popular Politics: The American North, 1865–1928* (New York: Oxford University Press, 1986), 5–10, 40–46, 149–150, 185–187; Robert Westbrook, "Politics as Consumption: Managing the Modern American Election" in *The Culture of Consumption: Critical Essays in American History, 1880–1980*, Richard Wightman Fox and T. J. Jackson Lears, eds. (New York: Pantheon Books, 1983), 152; Gabriel Kolko, *The Triumph of Conservatism: A Reinterpretation of American History, 1900–1916* (New York: Free Press of Glencoe, 1963), 2–6, 278–287; Alan Dawley, *Struggles for Justice: Social Responsibility and the Liberal State* (Cambridge, MA: Belknap Press of Harvard University Press, 1991), 3, 5, 134–138, 148–150; Arthur Link and Richard McCormick, *Progressivism* (Wheeling, IL: Harlan Davidson, 1983), 63–66; Robert Wiebe, *The Search for Order, 1877–1920* (New York: Hill and Wang, 1967), xiii–xiv, 12, 44–52. In the wake of Roosevelt's fireside chats, voter participation climbed to over 61 percent for the 1936 and 1940 presidential elections. It would be overstating the matter to attribute this turnaround only to Roosevelt's radio appearances, but radio's role cannot be discounted either.

3. David Thelen astutely discusses the ways that television audiences in the 1980s used the medium to help them enter into conversations directly with elected officials. In fact, such use of the mass media to create seemingly personal relationships within the realm of politics predates television. See David Thelen, *Becoming Citizens in the Age of Television: How Americans Challenged the Media and Seized Political Initiative during the Iran-Contra Debate* (Chicago: University of Chicago Press, 1996).

4. Betty Houchin Winfield, *FDR and the News Media* (Urbana: University of Illinois Press, 1990), 22, 104, 131; Steven Schoenherr, "Selling the New Deal: Stephen T. Early's Role as Press Secretary to Franklin D. Roosevelt" (Dissertation, University of Delaware, 1976), 38, 113; Hadley Cantril and Gordon Allport, *The Psychology of Radio* (New York: Arno Press, 1971; orig. 1935), 40; Edward Chester, *Radio, Television, and American Politics* (New York: Sheed and Ward, 1969), 31.

5. Winfield, 17–18, 104–105; Franklin Roosevelt, March 12, 1933, radio address, quoted in Franklin Delano Roosevelt, *Fireside Chats* (New York: Penguin Books, 1995), 1; Schoenherr, 133. There was no formal system for classifying Roosevelt's radio addresses, but in Schoenherr's judgment only thirteen of Roosevelt's talks prior to the outbreak of World War II in 1939 should be considered fireside chats.

6. Schoenherr, 41, 56, 70; Winfield, 2, 118, 234; Graham White, *FDR and the Press* (Chicago: University of Chicago Press, 1979), 154–158; Richard Steele, *Propaganda in an Open Society: The Roosevelt Administration and the Media, 1933–1941* (Westport, CT: Greenwood Press, 1985), 6–7, 11–16.

7. Stephen Early to Russell Burkhard, Aug. 5, 1933, Box 1, 1933 Folder, OF 136; Stephen Early to Marvin McIntyre, Oct. 26, 1933, Box 1, 1933 Folder, OF 136; William Hassett to Clifford Hunt, June 30, 1936, Box 1, Jan. to June 1936 Folder, OF 136; Stephen Early to Marvin McIntyre, Oct. 14, 1936, Box 2, July to Dec. 1936 Folder, OF 136; Stephen Early to A. T. Malmed, March 17, 1933, Box 5, Jan. to June 1933 Folder, OF 136; Merlin Aylesworth, New York, to Franklin Roosevelt, May 10, 1933, PPF 477; Franklin Roosevelt to Frank Walker, Feb. 13, 1936, PPF 1126; all in FDR Library. Winfield, 105, 108; Schoenherr, 112–115. For more on Roosevelt's use of radio as a political weapon, see Robert Brown, *Manipulating the Ether: The Power of Broadcast Radio in Thirties America*

(Jefferson, NC: McFarland and Company, 1998), 1, 9–11, 71. Brown overstates the influence of radio on politics in the 1930s, but he usefully discusses the political importance of controlling public information and setting a national agenda, and Roosevelt's success using radio for these goals.

8. Robert McElvaine, *The Great Depression: America, 1929–1941* (New York: Times Books, 1993), 367; Brown, 60; S. A. Harris, Moultrie, Georgia, to Franklin Roosevelt, Mar. 31, 1933, Box 5, Jan. to June 1933 Folder, OF 136. FDR Library; Schoenherr, 123–124, 137; Winfield, 109; Erik Barnouw, *A History of Broadcasting in the United States,* vol. 2, *The Golden Web: 1933–1953* (New York: Oxford University Press, 1968), 152. Roosevelt received an average of half a million letters in response to each of his radio addresses, according to historian Robert Brown. Over the course of the early years of the Depression, Roosevelt annually received about 16 letters for every 1,000 literate adults in the United States, Robert McElvaine reports—a figure that far surpassed the 4.7 letters per 1,000 literate adults Wilson received during World War I. Where Herbert Hoover's White House mailroom had a staff of one, Roosevelt's employed seventy people. David Kennedy, *Freedom from Fear: The American People in Depression and War, 1929–1945* (New York: Oxford University Press, 1999), 137. A sample of White House mail from one week in 1934 revealed the diversity of those letter writers: 46 percent of that week's mail came from laborers, 20 percent from businessmen and professionals, 15 percent from farmers, and 14 percent from clerks, according to McElvaine.

9. Schoenherr, 67–68, 77–78, 176–180; Steele, 22–23; Jeanette Sayre, *An Analysis of the Radiobroadcasting Activities of Federal Agencies* (Cambridge, MA: Radiobroadcasting Research Project, 1941), 25; Harold Ickes to Franklin Roosevelt, Sept. 28, 1939, Box 13, 1939 Folder, OF 6g, FDR Library.

10. *New York Herald-Tribune* in Sayre, 17; see also Sayre, 16–19, 28, 91, 110–113. Former president Herbert Hoover similarly charged that Roosevelt had made appointments to radio's licensing body, the Federal Communications Commission, based on political interests rather than technical expertise. Herbert Hoover, "The Reminiscences of Herbert Clark Hoover," interview by Allan Nevins and Frank Hill, Nov. 14, 1950, Columbia University, Oral History Project, Radio Unit, 20. Of course, both Democrats and Republicans liked to claim that their appointments to the commission were made without political motives. Regardless, networks did work to keep strong personal ties with the White House. CBS, for instance, appointed a friend of Roosevelt's, Henry Bellows, to head the company's interactions with the federal government. And the Mutual network—the third, much smaller, network in the 1930s—offered one of Roosevelt's sons a summer job. But not all network executives felt so friendly toward Roosevelt. Although he later behaved cordially, NBC president Merlin Aylesworth so hated Roosevelt that he once promised to leave the country if the then governor of New York were elected president of the United States. Steele, 19–20, 24–25, 128–132; Erik Barnouw, *A History of Broadcasting in the United States,* vol. 1, *A Tower in Babel; to 1933,* (New York: Oxford University Press, 1966), 273; Brown, 14–15, 33.

11. Chester, 33, 37; Barnouw, 52.

12. Roosevelt felt so comfortable with the broadcasting system that existed in the United States that he often had network executives help write his comments when he was to address the state of radio. In general, Roosevelt had little need to push institutional changes, as stations and networks tried to accommodate Roosevelt's wishes when possible to avoid provoking reform in the realm of radio. For more on Roosevelt's views

on the broadcasting system, see Franklin Roosevelt to Neville Miller, Nov. 25, 1940, Radio Broadcasting Folder, PPF 75; Franklin Roosevelt to James Cox, Jan. 27, 1940, 1940–1945 Folder, PPF 53; Franklin Roosevelt to Sol Taishoff, June 16, 1938, PPF 853; Franklin Roosevelt to Alfred McCosker, Sept. 14, 1934, in McCosker, Press Release, May 28, 1935, Box 1, 1935 Folder, OF 136, FDR Library; Roosevelt, April 12, 1940, in *The Complete Presidential Press Conferences of Franklin D. Roosevelt*, vol. 15 (New York: Da Capo Press, 1972), 253; Stephen Early, "Radio and Its Relation to Government" (address to the National Association of Broadcasters, July 11, 1939), PPF 853, FDR Library; Schoenherr, 76; Raymond McChesney, *Telecommunications, Mass Media, and Democracy: The Battle for the Control of U.S. Broadcasting, 1928–1935* (New York: Oxford University Press, 1993), 3, 183; Philip Rosen, *The Modern Stentors: Radio Broadcasters and the Federal Government* (Westport, CT: Greenwood Press, 1980), 178–179; Steele, 17–20.

13. Winfield, 1–2; White, ix; Steele, ix–x; Stephen Early to Vera Ingersoll, Nov. 21, 1934, Radio Broadcasting Folder, PPF 75, FDR Library.

14. Historians of Roosevelt and radio tend to focus their attention on his radio style. It deserves the attention—and more. As we shall see, though, it was not the only aspect of the president's on-air performances worth studying. For examples from a Roosevelt scholar and from a radio expert, see William Leuchtenburg, *Franklin D. Roosevelt and the New Deal, 1932–1940* (New York: Harper and Row, 1963), 330–331; and Barnouw, *The Golden Web*, 7–8.

15. Edward Deininger, Reading, Pennsylvania, to Franklin Roosevelt, Mar. 13, 1933, Box 9, PPF 200, FDR Library; F. B. Graham, Dubuque, Iowa, to Franklin Roosevelt, Mar. 13, 1933, Box 9, PPF 200, FDR Library. Graham was just one of the many listeners who understood Roosevelt's broadcasts in such personal terms. Even the *New York Times* agreed, saying Roosevelt's address "was like that of a man quietly sitting down with his neighbors. . . ." "Mr. Roosevelt's Address," *New York Times*, Apr. 30, 1935, 16. Television often gets cited as transforming politics in this fashion: historian Robert Westbrook has suggested that television revolutionized politics in the 1950s by creating an intimate and informal relationship between politicians and citizens. In fact, although Westbrook does not address it, radio deserves the credit for pushing such dramatic changes in politics roughly twenty years before the rise of television. See Westbrook, 155.

16. Franklin Roosevelt in Schoenherr, 110; Frances Perkins in Leuchtenburg, 330.

17. Henry Bellows to Stephen Early, Mar. 14, 1933, Box 5, Jan. to June 1933 Folder, OF 136, FDR Library; Will Rogers in Arthur Wertheim, *Radio Comedy* (New York: Oxford University Press, 1979), 79; Schoenherr, 111. Seventy percent of the words in Roosevelt's chats came from the five hundred most common English words.

18. E. B. Burkholder, St. Louis, to Franklin Roosevelt, Mar. 14, 1933, Box 9, PPF 200; James Dunn, Chicago, to Franklin Roosevelt, May 16, 1933, Box 12, PPF 200; J. C. Cassell, Mt. Joy, Pennsylvania, to Franklin Roosevelt, May 13, 1933, Box 12, PPF 200, FDR Library; for other examples see, L. C. Barthelmek, Ambridge, Pennsylvania, to Franklin Roosevelt, Mar. 13, 1933; H. L. Boyer, Geneva, New York, to Franklin Roosevelt, Mar. 14, 1933; Frederic Drake, New York, to Franklin Roosevelt, Mar. 13, 1933; all in Box 9, PPF 200, FDR Library.

19. Ernest Bormann, "A Rhetorical Analysis of the National Radio Broadcasts of Senator Huey P. Long" (Dissertation, State University of Iowa, 1953), 22, 75, 122, 244, 355, 380; Raymond Gram Swing, *Forerunners of American Fascism* (Freeport, NY: Books for Libraries Press, 1969; orig. 1935), 95; Morris Novik, Statement on the seventeenth

anniversary of LaGuardia's death, Sept. 20, 1964, Box 23A1, Folder 3, Morris Novik Donation, Fiorello LaGuardia Collection, LaGuardia and Wagner Archives, LaGuardia Community College, New York.

20. Frederic Drake, New York, to Franklin Roosevelt, Mar. 13, 1933, Box 9, PPF 200, FDR Library; Roy Crawford, Evanston, Illinois, to Franklin Roosevelt, Mar. 16, 1933, Box 9, PPF 200, FDR Library.

21. John Dos Passos, "The Radio Voice," *Common Sense* 3, no. 2 (Feb. 1934): 17. For a discussion of the film *Gabriel over the White House*, see Lawrence Levine, *The Unpredictable Past: Explorations in American Cultural History* (New York: Oxford University Press, 1993), 235–238.

22. "Precedent Answers the President," *Chicago Tribune*, June 30, 1934, 8; see also John Boettiger, "President Hits at Critics," *Chicago Tribune*, June 29, 1934, 1. Even the more moderate *New York Times*—which often expressed its faith in Roosevelt and his ability to use radio wisely—admitted that because radio offered skilled politicians the means to rouse their listeners' emotions, the medium might be used by unscrupulous leaders to manipulate the public; see "No Hasty Inflation," *New York Times*, May 9, 1933, 16. For Hoover's views on Roosevelt's use of radio, see Herbert Hoover, "Free Speech and Free Press," Nov. 8, 1937, address; and "Morals in Government," Apr. 26, 1938, address, in *Addresses upon the American Road* (New York: Charles Scribner's Sons, 1938), 277, 279, 340–341; Herbert Hoover, *The Challenge to Liberty* (New York: Charles Scribner's Sons, 1934), 136; Herbert Hoover, *The Memoirs of Herbert Hoover: The Great Depression 1929–1941*, vol. 3 (New York: Macmillan Company, 1952), 376; Hoover, "Reminiscences," 17. The jab at Roosevelt as a "radio crooner" came not from Landon himself, but a campaign worker in Minneapolis who introduced Landon to the crowd as the candidate who was not a "radio crooner." Landon, however, did consciously hope to take advantage of the difference between Roosevelt's polished radio style and his own awkward one. As he told a member of the New Deal's Brain Trust, Raymond Moley, shortly after the 1936 election, "I thought there might be some advantage in the antithesis between our deliveries. I guess I got too much antithesis." Alf Landon to Charles Fowler, Nov. 21, 1967, State Historical Society of Wisconsin, Madison, Wisconsin; Arthur Schlesinger, Jr., *The Age of Roosevelt: The Politics of Upheaval*, vol. 3 (Boston: Houghton Mifflin Company, 1960), 612.

23. Henry Blagden, Saranac, New York, to Franklin Roosevelt, Mar. 13, 1933, Box 9; Raymond Blatchley, Indianapolis, to Franklin Roosevelt, Mar. 13, 1933, Box 9; Mildred Stillman, New York, to Franklin Roosevelt, Sept. 15, 1939, Box 56; all in PPF 200, FDR Library.

24. "Mr. Roosevelt's Address," *New York Times*, Apr. 30, 1935, 16; Leuchtenburg 42–45; McElvaine, 141.

25. Eben Carey, Chicago, to Franklin Roosevelt, Mar. 11, 1933, Box 9; George Ludwig, Chambersburg, Pennsylvania, to Franklin Roosevelt, May 2, 1935, Box 21; both in PPF 200, FDR Library.

26. Reese Farnell, Talladega, Alabama, to Franklin Roosevelt, Oct. 13, 1937, Box 41, PPF 200, FDR Library.

27. Boyer to Roosevelt; Bradford Cadmus, New Canaan, Connecticut, to Franklin Roosevelt; John Dolan, Lowell, Massachusetts, to Franklin Roosevelt, March 14, 1933; both in Box 9, PPF 200, FDR Library.

28. Myra Gardner, Detroit, to Franklin Roosevelt, Mar. 12, 1933, Box 9; W. Hamilton Lee, Avondale Estates, Georgia, to Franklin Roosevelt, Apr. 29, 1935, Box 21; Harry Goldman, Los Angeles, to Franklin Roosevelt, Mar. 12, 1933, Box 9; all in PPF 200, FDR Library.

29. White, 69–70, 93, 129–134, 148–151; Winfield, 127, 144, 146; Steele, 127. In 1932, in the face of Hoover's failed presidency, 52 percent of the press supported the incumbent; by 1940, 64 percent of the press staked out positions opposing Roosevelt. This made it easy for Roosevelt to point to the differences between newspaper endorsements and election returns as evidence that newspapers were out of step with popular opinion and that his government better represented the people.

30. Roosevelt in Winfield, 104; Harry Butcher to Marvin McIntyre, Nov. 11, 1936, 1936–1939 Folder, OF 256, FDR Library; Daniel Roper to Franklin Roosevelt, Apr. 18, 1938, 1938 Folder, OF 340, FDR Library. Admittedly, Butcher's view was somewhat self-serving: as a CBS executive, he had an interest in making Roosevelt feel grateful toward radio. Nonetheless, Butcher's point resonated among some Roosevelt advisors. Eleanor Roosevelt, for instance, suggested to her husband that he should take to the air more often by broadcasting his twice-weekly press conferences. It would allow him to circumvent the reporters at those conferences and reach the people directly, she hinted. Bess Furman to Eleanor Roosevelt, no date, Box 2, July to Dec. 1941 Folder, OF 36, FDR Library.

31. Franklin Roosevelt to Merlin Aylesworth, May 18, 1933, 1933–1934 Folder, OF 228, FDR Library; see also Roosevelt to Miller; and Franklin Roosevelt to Merlin Aylesworth, Nov. 8, 1933, Radio Broadcasting Folder, PPF 75, FDR Library; Herbert Hoover to James Aitken, Sept. 16, 1931, in Hoover, *Public Papers of the Presidents of the United States*, vol. 1931 (Washington, DC: United States Government Printing Office, 1976), 439; Winfield, 2, 18.

32. John Studebaker, "Proposals, Involving Education, Which Are Designed to Restore and Increase Confidence in the Admin.," Box 12, Sept. 1935 Folder, OF 6g; John Studebaker, "An Educational Approach to Government at Work by Radio and Other Series of Educational Broadcasts," Box 13, 1937 Folder, OF 6g; Harold Ickes to Franklin Roosevelt, Sept. 28, 1939, Box 13, 1939 Folder, OF 6g; all in FDR Library.

33. Franklin Roosevelt to Lennox Lohr, July 22, 1937, Radio Broadcasting Folder, PPF 75, FDR Library; Roosevelt to Miller; Studebaker, "An Educational Approach to Government at Work by Radio and Other Series of Educational Broadcasts," Box 13, 1937 Folder, OF 6g, FDR Library, 1; Stephen Early, interview by George Homes, transcript, July 6, 1934, NBC, 1933–1934 Folder, OF 228, FDR Library; Franklin Roosevelt to Anning Prall, Nov. 6, 1936, PPF 477; Franklin Roosevelt to William Paley, Apr. 22, 1938, PPF 984, FDR Library; Winfield, 103; White, 129–130, 157; Steele, 6.

34. Fiorello LaGuardia to Elliot Sharp, May 11, 1938, Box 3222, Folder 5, LaGuardia Dept. Correspondence, New York Municipal Archives, New York; Fiorello LaGuardia, transcript of testimony before FCC, Mar. 28, 1940, Box 23A1, Folder 6, Morris Novik Donation, Fiorello LaGuardia Collection, 6–18; Fiorello LaGuardia, statement on WNYC, no date, Box 3244, Folder 6, LaGuardia Dept. Correspondence; Morris Novik, "Radio and Labor," no date, Box 23A1, Folder 2, Novik Donation, LaGuardia Collection. It is worth noting that while both LaGuardia and Roosevelt believed in using broadcasting to distribute civic information to the public, they disagreed on the question of public versus private ownership of radio. Roosevelt accepted commercial control of the air,

whereas LaGuardia favored giving publicly owned stations equal access to the air. Indeed, LaGuardia took his crusade for public broadcasting to Washington, appearing before the FCC on behalf of WNYC: "Any publicly owned station operated by a state or a subdivision of a state, operated on a noncommercial basis for governmental purposes, education and recreation, has priority over a commercial station operated for profit," he argued.

35. Herbert Hoover to H. P. Davis, Nov. 1, 1930, in Herbert Hoover, *Public Papers of the Presidents of the United States*, vol. 1930 (Washington, DC: United States Government Printing Office, 1976), 462; Hoover to James Aitken, in *Public Papers of the Presidents of the United States*, vol. 1931 (Washington, DC: United States Government Printing Office,1976), 439; Herbert Hoover, "Morals in Government," Sept. 28, 1938, address, in *Further Addresses upon the American Road* (New York: Charles Scribner's Sons, 1940), 16–17; Hoover, *Memoirs*, 453; Chester, 28; Sayre, 28, 91, 110–113. In 1940 Congress eliminated funding for the Office of Education's radio activities. Further cutbacks might have followed shortly if the United States had not entered World War II, leading Congress to accept government-sponsored broadcasting for the war effort. For Early's rebuttal, see Early interview.

36. By the Vietnam War, for instance, the press on its own proved reluctant to challenge the federal government's statements directly. Only when there was dissent within the government or no version of events coming from the government would the press lay out an adversarial position. For useful discussions of the influence of objectivity on the press in Vietnam see Daniel Hallin, *The "Uncensored War": The Media and Vietnam* (Berkeley: University of California Press, 1986); Clarence Wyatt, *Paper Soldiers: The American Press and the Vietnam War* (Chicago: University of Chicago Press, 1995).

37. Hoover, *Addresses upon the American Road*, 277, 279, 340; Hoover, *Further Addresses upon the American Road*, 16; Hoover, *Memoirs*, 219, 376; Hoover, *The Challenge to Liberty*, 17. No doubt Hoover mounted such vitriolic attacks on Roosevelt's use of radio in part because he felt the Democrats had used the medium to blast him personally. But the worries he raised were genuine. After Republicans recaptured the presidency, Hoover mellowed his critique, but he still believed the medium posed problems as an instrument of propaganda. Hoover "Reminiscences," 16–17.

38. Orville Brown, Phoenix, to Franklin Roosevelt, Box 9; Laura Lunde, Chicago, to Franklin Roosevelt, Oct. 11, 1937, Box 41; both in PPF 200, FDR Library; Dolan to Roosevelt.

39. Dolan to Roosevelt. R. J. Barlet, San Francisco, to Franklin Roosevelt, March 13, 1933, Box 9; Marvin Coontz, Riverside, California, to Franklin Roosevelt, March 13, 1933, Box 9; Queening de Mena, Pasadena, California, Franklin Roosevelt, Box 12; all in PPF 200, FDR Library.

40. Oliver Chryers, St. Paul, to Franklin Roosevelt, Oct. 1, 1934, Box 18, PPF 200, FDR Library.

41. Margaret Moore, Normal, Illinois, to Franklin Roosevelt, Sept. 6, 1938, Box 35, PPF 200, FDR Library. Similarly, the *New York Times* suggested that Roosevelt's use of radio gave "a new meaning to the old phrase about a public man, 'going to the country.'" "Banking Normal Again," *New York Times*, Mar. 14, 1933, 14. Historians like Leuchtenburg and Robert McElvaine are right that popular participation in national politics was a crucial legacy of the New Deal. Leuchtenburg, 331; McElvaine, 336. But they, like many political historians, understate the role radio played in that transformation. Many letters to Roosevelt attacked the workings of the New Deal on the local level while still praising the president. See, for instance, McElvaine, 113. Clearly, then, the decisions letter writers

made about the head of their national government were based on something *other* than the actual policies in action.

42. Anne Clapp, Duluth, Minnesota, to Franklin Roosevelt, May 8, 1933, Box 12; Newton Fetter, Cambridge, Massachusetts, to Franklin Roosevelt, Mar. 15, 1933, Box 9; both from PPF 200, FDR Library.

43. Dunn to Roosevelt.

44. Ralph Coryell, Birmingham, Michigan, to Franklin Roosevelt, Sept. 30, 1934, Box 18; Mrs. Nellie Lock, Enid, Oklahoma, to Franklin Roosevelt, Apr. 30, 1935, Box 21; James Cullen, San Francisco, to Franklin Roosevelt, Oct. 3, 1934, Box 18; all in PPF 200, FDR Library.

45. George Allen, Richmond, Virginia, to Franklin Roosevelt, Apr. 14, 1938, Box 50, Con A–F Folder, PPF 200, FDR Library.

46. Paul Stern, New York, to Franklin Roosevelt, Apr. 23, 1938, Box 50, Pro S Folder, PPF 200, FDR Library.

47. Francis Perkins in Steele, 7. James Fisher, Robesonia, Pennsylvania, to Franklin Roosevelt, May 8, 1933, Box 12; Oscar Dallia, Chicago, to Franklin Roosevelt, Mar. 13, 1933, Box 9; both in PPF 200, FDR Library.

48. Roosevelt to Miller.

49. The decline of public forms of civic involvement in the latter portion of the twentieth century has been attributed to many factors. In his thorough study of the subject, political scientist Robert Putnam concluded television is the chief culprit. Television, he says, "privatizes our leisure time," replacing group activities with private ones. It is, he says, the generation that grew up in the age of television that eschews civic engagement; the technology, he concludes, "may indeed be undermining our connections with one another and with our communities." As often happens, he overlooks the role television's predecessor, radio, played in this process. It was radio that first changed our ideas of civic participation, paving the way for the disengagement Putnam reports. If we want to understand the foundations of the political problem Putnam explores, we need to look earlier in time than the arrival of television. See Robert Putnam, "The Strange Disappearance of Civic America," *American Prospect* 24 (Winter 1996): 44–48.

50. Bryan Doble, Hinsdale, Illinois, to Franklin Roosevelt, Apr. 15, 1938, Box 50, Con A–F Folder, PPF 200, FDR Library.

51. Benjamin Weis, Columbus, Ohio, to Franklin Roosevelt, Oct. 14, 1937, Box 41; for other examples, see W. A. Endy, Chicago, to Franklin Roosevelt, no date, Box 9; Bernard Sachs, New York, to Franklin Roosevelt, Apr. 25, 1938, Box 50, Pro S Folder; all in PPF 200, FDR Library.

52. Fiorello LaGuardia to James McDonald, July 13, 1939, Box 3244, Folder 6, LaGuardia Dept. Correspondence, New York Municipal Archives.

53. Dunn to Roosevelt.

54. Arthur Schlesinger, Cambridge, Massachusetts, to Franklin Roosevelt, May 8, 1935, PPF 2501, FDR Library.

55. Schlesinger to Roosevelt.

Chapter 4

1. Long, of course, is only the most obvious example of another politician who could be included in this group. Smith was never the radio personality that Coughlin was,

but he periodically took his religious persona to the air to further his political agenda. McPherson, who offered both religious salvation and through it medical salves, primarily flourished in the 1920s, before tapping a truly mass audience was possible. Baker never had Brinkley's following or political power, but he too offered medical salvation—a cure for cancer—through the airwaves. See, for examples, Alan Brinkley, *Voices of Protest: Huey Long, Father Coughlin, and the Great Depression* (New York: Vintage Books, 1983), 62, 71, 143–144, 169; Leo Ribuffo, *The Old Christian Right: The Protestant Far Right from the Great Depression to the Cold War* (Philadelphia: Temple University Press, 1983), 149–150; Lynn Dumenil, *Modern Temper: American Culture and Society in the 1920s* (New York: Hill and Wang, 1995), 178–181; Gerald Carson, *The Roguish World of Doctor Brinkley* (New York: Rinehart and Company, 1960), 3–5.

2. Franklin Roosevelt quoted in Donald Warren, *Radio Priest: Charles Coughlin, the Father of Hate Radio* (New York: Free Press, 1996), 64–65.

3. Steven Schoenherr, "Selling the New Deal: Stephen T. Early's Role as Press Secretary to Franklin D. Roosevelt" (Dissertation, University of Delaware, 1976), 123–124, 137; Erik Barnouw, *A History of Broadcasting in the United States* vol. 2, *The Golden Web: 1933–1953* (New York: Oxford University Press, 1968), 152; Betty Houchin Winfield, *FDR and the News Media* (Urbana: University of Illinois Press, 1990), 109; *Fortune* quoted in David Bennett, *Demagogues in the Depression: American Radicals and the Union Party, 1932–1936* (New Brunswick, NJ: Rutgers University Press, 1969), 54, see also 44; Brinkley 83, 119; Robert McElvaine, *The Great Depression: America, 1929–1941* (New York: Times Books, 1993), 238; Carson, 143.

4. R. Rohrbough, Houston, to Franklin Roosevelt, Apr. 29, 1935, Box 21; H. L. Boyer, Geneva, New York, to Franklin Roosevelt, Mar. 14, 1933, Box 9; both in PPF 200, Franklin D. Roosevelt Library, Hyde Park, New York (hereafter FDR Library).

5. W. D. Dixon, Springfield, Massachusetts, to Franklin Roosevelt, May 8, 1933, Box 12; Margaret Englemann, Portland, to Franklin Roosevelt, May 7, 1933, Box 12; Kate Carmichael, Goldsboro, North Carolina, to Franklin Roosevelt, Apr. 2, 1933, Box 9; all in PPF 200, FDR Library. See also McElvaine, 112–117; William Leuchtenburg, *Franklin D. Roosevelt and the New Deal, 1932–1940* (New York: Harper and Row, Publishers, 1963), 331; Richard Steele, *Propaganda in an Open Society: The Roosevelt Administration and the Media, 1933–1941* (Westport, CT: Greenwood Press, 1985), 7; Robert Brown, *Manipulating the Ether: The Power of Broadcast Radio in Thirties America* (Jefferson, NC: McFarland and Company, 1998), 9–12, 25–27.

6. For a sense of the range of views of Coughlin—some of which overlap—see Brinkley, 143–148, 157–164; McElvaine, 237–240, 248–249; Bennett, 56, 294; Warren, 28; Michael Kazin, *The Populist Persuasion: An American History* (New York: Basic Books, 1995), 109–133; Raymond Gram Swing, *Forerunners of American Fascism* (Freeport, NY: Books for Libraries Press, 1969; orig. 1935), 22–59. On Dr. Brinkley, Ansel Resler, Erik Barnouw, and Gene Fowler and Bill Crawford largely see him as a radio pioneer, while Gerald Carson focuses on him as a theatrical medical fraud. Indeed, Brinkley was a significant radio innovator. In the 1920s, the doctor was one of the first popular broadcasters to speak informally and intimately on the air. By the 1930s, an array of radio speakers, including Roosevelt, used some sort of personal speaking style with mixed degrees of success on the air. More specifically, Barnouw suggests that Huey Long's distinctive colloquial intimacy followed the stylistic path Brinkley paved. Ansel Harlan Resler, "The Impact of John R. Brinkley on Broadcasting in the United States" (Dissertation,

Northwestern University, 1958), 248–257; Erik Barnouw, *A History of Broadcasting in the United States,* vol. 1, *A Tower in Babel; to 1933,* (New York: Oxford University Press, 1966), 168–172; Gene Fowler and Bill Crawford, *Border Radio: Quacks, Yodelers, Pitchmen, Psychics, and Other Amazing Broadcasters of the American Airwaves* (Austin: University of Texas Press, 2002), 18; Carson 3–8.

7. There are, of course, good reasons the duo are rarely paired. Brinkley appealed to an overwhelmingly rural and Protestant audience from the South and plains states; Coughlin's biggest fans tended to be urban, often Catholic, listeners from the North and Midwest. Moreover, while Brinkley entered politics periodically, he tended to focus on state and local races and did not make politics central to his radio presence. Coughlin, on the other hand, fashioned himself into a force in national politics through his constant broadcasting efforts to shape the nation's agenda.

8. Swing, 34; Brinkley, 89–101; Bennett, 33–37; McElvaine, 238; Kazin, 113.

9. Swing, 40; Warren, 34, 291–292; Bennett, 34–36, 58; Brinkley, 83, 119–120. Whether for his own gain or for the cause he led, Coughlin occasionally sought to use his political influence for financial profit. In 1934 he ardently urged the United States to remonetize silver. It turned out his personal secretary had bought Coughlin's organization futures contracts for over fifteen tons of silver. Brinkley, 124–125.

10. Ernie Pyle quoted in Resler, 4.

11. For Brinkley's early life, see Carson, 12–40; Resler, 9; Francis Schruben, *Kansas in Turmoil, 1930–1936* (Columbia: University of Missouri Press, 1969), 28; Friends of Democracy memo to FCC, Aug. 29, 1940, Box 193, General Correspondence 1927–1946, Office of the Executive Director, FCC Collection, Research Group 173, National Archives, College Park, Maryland (hereafter "General Correspondence, FCC"); Clement Wood, *The Life of a Man: A Biography of John R. Brinkley* (Kansas City: Goshorn Publishing Company, 1934). Note that Brinkley commissioned Wood to write *The Life of a Man*; the book essentially tells Brinkley's story as he wished it told for publicity purposes.

12. Carson, 87–91, 99–101; Resler, 80–82; Barnouw, *A Tower in Babel,* 170–171; Schruben, 29; Wood, 216–220. Some years later, an analysis of one of Brinkley's medicines for recovering surgery patients determined that the liquid contained only water, dye, and a trace amount of acid. Each vial cost the doctor eighteen cents and sold for a hundred dollars. Carson, 193.

13. C. P. Collins, Bellefontaine, Ohio, to Franklin Roosevelt, Feb. 15, 1934, Box 193, General Correspondence, FCC; Carson, 153, 177, 181, 186; Resler, 82, 125, 165, 207, 243; Schruben, 82–83; Wood, 216–220; Fowler and Crawford, 25, 36. Through the 1930s, the FCC received letters about Brinkley from listeners in states from Rhode Island to Wisconsin to California. For examples, see Dr. Brinkley Folder, Box 192, and Box 193, General Correspondence, FCC.

14. Janet Bonthuis to *Vox Pop,* Nov. 20, 1939, Folder 7, Box 1, Series I, Vox Pop Collection, Broadcast Pioneers Library of American Broadcasting, University of Maryland, College Park, MD.

15. Bernard Sachs, New York, to Franklin Roosevelt, Apr. 25, 1938, Box 50, Pro S Folder; W. A. Endy, Chicago, to Franklin Roosevelt, no date, Box 9; Benjamin Weis, Columbus, Ohio, to Franklin Roosevelt, Oct. 14, 1937, Box 41; all in PPF 200, FDR Library.

16. John Dos Passos, "The Radio Voice," *Common Sense* 3, no. 2 (Feb. 1934): 17.

17. Franklin Roosevelt, "First Inaugural Address," American Rhetorical Discourse, 2nd ed., ed. Ronald Reid (Prospect Heights, IL: Waveland Press, 1995), 723; McElvaine, 118, 316, 325; Graham White, *FDR and the Press* (Chicago: University of Chicago Press, 1979), 130; Franklin Delano Roosevelt, *Fireside Chats* (New York: Penguin Books, 1995), 1–8. Note Roosevelt's bold reversal in the fireside chat statement. Instead of telling listeners "your burdens are mine" as one might have expected from a leader seeking popular approval, Roosevelt confidently declared that his burdens were shared by his listeners. Instead of asking listeners to let him represent them, he told listeners he represented them perhaps better than they represented themselves.

18. Mary O'Gorman, Toledo, Ohio, to Federal Radio Commission, Mar. 29, 1933; Dorothy Parker, Highland Park, Michigan, to Federal Radio Commission, Mar. 29, 1933; both in Coughlin Folder, Box 199, General Correspondence, FCC. My study considers letters sent to the Federal Communications Commission. Historian David Bennett found a similar phenomenon in the letters Coughlin's supporters sent directly to his critics. Bennett, 62.

19. Charles Coughlin quoted in Swing, 44–46, 56; Hadley Cantril and Gordon Allport, *The Psychology of Radio* (New York: Arno Press, 1971; orig. 1935), 8; Bennett, 42.

20. O'Gorman to Federal Radio Commission; Marion Orr, Hamden, Connecticut, to NBC, Nov. 28, 1938, Box 200, General Correspondence, FCC.

21. Charles Coughlin quoted in Brinkley, 150; for a thorough exploration of Coughlin's thinking, see Brinkley, 143–168; Warren 28, 113; Bennett 58–61, 294–296; Wallace Stegner, "The Radio Priest and His Flock," *The Aspirin Age: 1919–1941*, Isabel ed. Leighton (New York: Simon and Schuster, 1949), 240–241.

22. Resler, 84, 94, 115, 205, 244; Schruben, 29; Barnouw, *A Tower of Babel*, 171, 258–259; Carson, 135–136, 146–152; Ethel Rauber, Newton, Kansas, to Franklin Roosevelt, Feb. 28, 1934, Box 193, General Correspondence, FCC. Some expressed the link between Brinkley and the common populace by suggesting that politicians who tried to shut down Brinkley's broadcasting would find themselves voted out of office. In 1931, for instance, Helen Smerchek warned Herbert Hoover's administration that closing the airwaves to Brinkley would hurt the administration in 1932. Helen Smerchek, Topeka, Kansas, to Federal Radio Commission, Oct. 3, 1931, Dr. Brinkley Folder, Box 192, General Correspondence, FCC.

23. Wood, 227, 271, 315; Schruben, 32–37; Resler, 112, 178.

24. Fred Reilly, Burlingame, California, to Federal Radio Commission, Nov. 19, 1931; E. A. Boehme, Offerle, Kansas, to Federal Radio Commission, Nov. 26, 1931; both in Dr. Brinkley Folder, Box 192, General Correspondence, FCC.

25. Wood, 202, 227–229, 247, 282; Resler, 112; Schruben, 45; Brinkley, 22.

26. Carson, 91; Brinkley mailing quoted, 94; Wood, 230.

27. See Carson, 163; Resler, 187.

28. F. B. Graham, Dubuque, Iowa, to Franklin Roosevelt, Mar. 13, 1933; Carl Geigand, Buffalo, to Franklin Roosevelt, Mar. 13, 1933; both in Box 9, PPF 200, FDR Library.

29. Jeffrey Tulis offers a useful discussion of the transformation of the role and authority of the president: in the nineteenth century, the president was primarily head of the government and not a popular leader; in the twentieth century, presidents increasingly came to see themselves as leaders of the people and came to rely on appealing to the people at large as a source of power. Tulis understates the importance of technologies of

communication in this process, but he lays out an important shift. See Jeffrey Tulis, *The Rhetorical Presidency* (Princeton, NJ: Princeton University Press, 1987), throughout, but especially 3–4, 13–17, 124–136.

30. On Coughlin and the World Court, see Brinkley, 134–137; Swing, 59; Warren, 63–65. It was the quest for this sort of power, some contemporaries and later critics argued, that motivated Coughlin's actions through the 1930s. Bennett, 56; Stegner, 244.

31. Quoted in Carson, 126; Resler, 165, 207; Barnouw, *A Tower of Babel*, 171–172; Schruben, 29; Friends of Democracy memo to FCC.

32. During the campaign Brinkley ran on a quintessential platform of promises. He promised to restore a connection between voters and their government by installing a radio microphone in the governor's office and house. He promised a tax-free medical clinic for the poor, free schoolbooks, free auto tags, and a better deal for working folks. He promised just about anything he could think of, right down to increasing rainfall on the prairie—a goal he planned to accomplish by building a lake in every county. Schruben, 31–41, 89–100; Carson, 156, 162, 165–173; Friends of Democracy memo to FCC.

33. Carson, 177, 183–187, 200–201; Resler, 131, 148–154, 165; Barnouw, *The Golden Web*, 130; Robert McChesney, *Telecommunications, Mass Media, and Democracy: The Battle for the Control of U.S. Broadcasting, 1928–1935* (New York: Oxford University Press, 1993), 29; Schruben, 141, 152. In Colorado, W. F. Wilcox took Brinkley's interference with a Denver station particularly hard: "It makes people so angry they feel like committing murder, having money invested in radios and unable to get the best service as our best station is killed," he wrote. W. F. Wilcox, Montrose, Colorado, to FCC, Feb. 5, 1936, Box 193, General Correspondence, FCC. The FCC file on Brinkley is larger than its file on any other radio figure in the 1930s save Coughlin: from across the country, listeners objected to the doctor drowning out local stations. Dr. Brinkley Folder, Box 192 and Box 193, General Correspondence, FCC.

34. John Boettiger, "President Hits at Critics," *Chicago Tribune*, June 29, 1934, 1; "Precedent Answers the President," *Chicago Tribune*, June 30, 1934, 8. Orro, "Trying to Hypnotize Uncle Again," editorial cartoon, *Chicago Tribune*, Oct. 6, 1936, Folder Oct. 1–16, 1936, Box 18; S. J. Ray, cartoon "Delivering a Message to Congress," Jan. 4, 1935, Box 13, Folder Jan. 1–15, 1935, both in Basil O'Connor Collection, FDR Library; Herbert Hoover, *Further Addresses upon the American Road* (New York: Charles Scribner's Sons, 1940), 16–17.

35. Franklin Roosevelt to Arthur Schlesinger, May 14, 1935, PPF 2501, FDR Library; Brown, 69, 96; Warren, 65. When Roosevelt's ambassador to Mexico, Josephus Daniels, attended the 1933 radio conference with the Mexican government, he focused much of his attention on Brinkley's Mexican station. "Above all else," the ambassador said, he hoped for an agreement between the two countries that would silence the likes of Brinkley. Brinkley was not the only problematic speaker broadcasting across the Rio Grande, but he was the biggest and most troubling from the administration's point of view. Josephus Daniels to Franklin Roosevelt, July 12, 1933, OF 525, FDR Library; Josephus Daniels to Franklin Roosevelt, July 29, 1933, Folder 1933–1934, OF 237, FDR Library.

36. Ernest Stillman, New York, to Franklin Roosevelt, Mar. 10, 1933, Coughlin Folder, Box 199, General Correspondence, FCC.

37. D. L. Browning, London, Ohio, to Federal Radio Commission, Nov. 9, 1931, Coughlin Folder, Box 199, General Correspondence, FCC.

38. William Wood, Baltimore, to Herbert Hoover, Nov. 13, 1931, Coughlin Folder, Box 199; R. B. Wood, Adams, Wisconsin, to FCC, Feb. 11, 1935, Coughlin Folder, Box 199; George Clarke, Waukegan, Illinois, to FCC, Nov. 3, 1935, Box 200; all in General Correspondence, FCC.

39. Clarke to FCC.

40. Geo Richards, Thurmond, West Virginia, to FCC, Feb. 24, 1939; Anonymous to Justice Department, April 1936 both in Box 193, General Correspondence, FCC.

41. Brinkley, 254–268; Kazin, 124–132; Stegner, 249; Warren, 215–247; McElvaine, 240; Warren, 95–96. In 1936 Coughlin's candidate, Lemke, won less than 2 percent of the vote nationwide. Coughlin had not promised victory, but he had vowed to retire from radio if he could not deliver nine million votes. Coughlin fell 8.1 million votes short of that goal. He was back on the air six weeks later.

42. Carson, 219, 237–250; Friends of Democracy memo to FCC; C. M. Willis, Kilgore, Texas, to Wright Patman, U.S. Representative, Dec. 19, 1937, Box 193, General Correspondence, FCC; Resler, 162, 166–169. Brinkley died the year after losing his access to the air.

43. Brinkley, 159, 168.

44. For more on the rising importance of image in the twentieth century, see Warren Susman, *Culture as History: The Transformation of American Society in the Twentieth Century* (New York: Pantheon Books, 1984), 165–166, 272–283; Daniel Boorstin, *The Image: A Guide to Pseudo-Events in America* (New York: Harper and Row, Publishers, 1961), 45–48, 57, 61, 246–250.

45. Stegner, 234–236; W. L. Thickstrum, Tullahoma, Tennessee, to Federal Radio Commission, Nov. 20, 1932, Dr. Brinkley Folder, Box 192, General Correspondence, FCC. As one of the many money-making schemes Brinkley aired, his station invited listeners to send in a dollar for on-air astrological advice. The Thickstrums sent in their question and money, but received no reply, eventually prompting them to look to the federal government for help.

In case the Depression and sickness were not enough to inspire desperation in his listeners, Brinkley stoked their fears. He played on their concerns that much in their lives was out of their hands and on their desires to find a measure of control. He cast his listeners' troubles in terms familiar to many of his Bible-belt followers. To be healed, he told listeners, one must first accept one's sickness. "Disease is working on you day and night," Brinkley announced. "You can feel those bacteria in the vicinity of the prostate itching and crawling. Yes you can tell it—you don't want to die. . . . Act promptly and don't let the disease get any more start on you. . . . You must come to the Brinkley hospital with a contrite heart, and be humble and want health. . . . You women try to get your husbands disgusted with their diseased condition." John Preston, radio inspector's report, Jan. 5, 1938, Box 193, General Correspondence, FCC.

46. Huey Long quoted in Ernest Bormann, "A Rhetorical Analysis of the National Radio Broadcasts of Senator Huey P. Long" (Dissertation, State University of Iowa, 1953), 492.

Chapter 5

1. Paul Lazarsfeld to Frank Stanton, July 14, 1938, Folder 6, Box 3A, Series I, Paul Lazarsfeld Collection, Butler Library, Columbia University, New York, NY (hereafter Lazarsfeld Collection).

2. Hadley Cantril and Gordon Allport, *The Psychology of Radio* (New York: Arno Press, 1971; orig. 1935), vii.

3. Dorothy Ross, *The Origins of American Social Science* (Cambridge: Cambridge University Press, 1991), xv, 53–64, 312–316, 369–370, 388–400, 406, 428–430, 474–475; John Diggins, *The Promise of Pragmatism: Modernism and the Crisis of Knowledge and Authority* (Chicago: University of Chicago Press, 1994), 7–11, 226–228.

4. Paul Lazarsfeld, "Remarks on Administrative and Critical Communications Research," *Studies in Philosophy and Social Science* 9, no. 1 (1941): 8–9.

5. John Peters argues that the debate between Lazarsfeld and Adorno was the central debate in communication theory of the twentieth century because it was, in fact, a debate over whether or not mass communication was even possible. John Durham Peters, *Speaking into the Air: A History of the Idea of Communication* (Chicago: University of Chicago Press, 1999), 223.

6. Ann Pasanella, *The Mind Traveller: A Guide to Paul F. Lazarsfeld's Communication Research Papers* (New York: Freedom Forum Media Studies Center, 1994), 19; for a summary of radio research at universities around the country in the late 1930s, see Isabelle Wagner, ed., "Current Radio Research in Universities," *Journal of Applied Psychology* 23, no. 1 (Feb. 1939): 192–206; for Lazarsfeld's own account of his institution-building efforts, see Paul Lazarsfeld, "An Episode in the History of Social Research: A Memoir," in *The Intellectual Migration: Europe and America, 1930–1960*, ed. Donald Fleming and Bernard Bailyn (Cambridge, MA: Belknap Press of Harvard University Press, 1969), 302.

7. Paul Lazarsfeld to Hadley Cantril, Aug. 8, 1937 (?), Folder 1, Box 27, Series I, Lazarsfeld Collection. For a detailed account of the general pursuit of empirical means of studying society, see Ross.

8. "The Essential Value of Radio to All Types of Listeners," Draft of proposal to found the Princeton Radio Research Project, Folder 7, Box 26, Series I, Lazarsfeld Collection, 1–3; "Proposal for Continuation of Radio Research Project ," Folder 14, Box 26, Series I, Lazarsfeld Collection, 1–2. For Lazarsfeld the alliance with the Rockefeller Foundation was a fairly natural one. The Rockefeller Foundation sought to promote the application of scientific research to social problems, using its grants to push many scholars down that path. For more on the Rockefeller Foundation's influence on social sciences, see Ross, 400–404.

9. Lazarsfeld, *Radio and the Printed Page: An Introduction to the Study of Radio and Its Role in the Communication of Ideas* (New York: Duell, Sloan, and Pearce, 1940), xi, 200.

10. Lazarsfeld, *Radio and the Printed Page*, xi–xii; "Proposal for Continuation of Radio Research Project."

11. Paul Lazarsfeld, "The Effects of Radio on Public Opinion," *Print, Radio, and Film in a Democracy: Ten Papers on the Administration of Mass Communications in the Public Interest; Read before the Sixth Annual Institute of the Graduate Library School, the University of Chicago, August 4–9, 1941*, ed. Douglas Waples (Chicago: University of Chicago Press, 1942), 76.

12. Cantril and Allport, 20, 27.

13. Cantril and Allport, 27, 258; Glenn Frank, "Radio as an Educational Force," *The Annals of the American Academy of Political and Social Sciences* 177 (Jan. 1935), *Radio: The Fifth Estate*, ed. Herman Hettinger, 119; Lazarsfeld, *Radio and the Printed Page*, xvi, 5–46, 148, 198.

14. Lazarsfeld, *Radio and the Printed Page*, 130–132; see also Edward Suchman, "Invitation to Music: A Study of the Creation of New Music Listeners by the Radio," *Radio Research 1941*, ed. Paul Lazarsfeld and Frank Stanton (New York: Duell, Sloan and Pearce, 1941), 142–188. For more on Victorian notions of culture, see Daniel Czitrom, *Media and the American Mind: From Morse to McLuhan* (Chapel Hill: University of North Carolina Press, 1982), 31–35; Henry May, *The End of American Innocence: A Study of the First Years of Our Own Time, 1912–1917* (Chicago: Quadrangle Books, 1959), 30–36; Lawrence Levine, *Highbrow/Lowbrow: The Emergence of Cultural Hierarchy in America* (Cambridge, MA: Harvard University Press, 1988), 200–231; Alan Trachtenberg, *The Incorporation of America: Culture and Society in the Gilded Age* (New York: Hill and Wang, 1982), 140–161.

15. Lazarsfeld, *Radio and the Printed Page*, 201; Cantril and Allport, 18, 259–260.

16. Cantril and Allport, 20, 24; Frank, 121; see also Charles Siepmann, "Radio and Education," *Studies in Philosophy and Social Science* 9, no. 1 (1941): 118. The academics who first began considering radio in the 1920s did not directly give rise to media studies as a field, but clearly some of their thinking echoed in the 1930s. For a discussion of those early students of radio, see Czitrom, 91–121.

17. "Proposal for Continuation of Radio Research Project," 2.

18. Cantril and Allport, vii–iii; "Proposal for Continuation of Radio Research Project"; Lazarsfeld, *Radio and the Printed Page*, xiv. A more concrete illustration of this pragmatic emphasis of ideas as actions can be seen in the many researchers who sought to understand more than simply the popularity of a program, but how listeners acted after listening to those programs. For example, writing for a Lazarsfeld-edited journal, Harry Link and Philip Corby declared that the proper measure of a program's effectiveness was how it impacted sales of the product it advertised. Harry Link and Philip Corby, "Studies in Radio Effectiveness by the Psychological Corporation," *Journal of Applied Psychology* 24, no. 6 (Dec. 1940). The question of the use of knowledge prodded diverse social scientists from the Progressive era onward: social research, they suggested, should not just explain, but serve society. Ross, 367–368, 400, 404.

19. Lazarsfeld, "Remarks on Administrative and Critical Communications Research," 2–3.

20. Lazarsfeld, *Radio and the Printed Page*, 331–333. For example, Lazarsfeld suggested that if suburbanization became the American norm, radio would become more crucial as families sought means to access news and entertainment from dispersed homes. But, he continued, if the population grew denser, radio would become less important: urban dwellers would have easy access to movies, evening schools, and the like, and would have little need for radio. See also Suchman, 188.

21. Lazarsfeld, *Radio and the Printed Page*, 133–134.

22. Cantril and Allport, vii; "The Essential Value of Radio to All Types of Listeners," 3; *Journal of Applied Psychology* 23, no. 1 (Feb. 1939); *Journal of Applied Psychology* 24, no. 6 (Dec. 1940); Paul Lazarsfeld, "Introduction by the Guest Editor," *Journal of Applied Psychology* 24, no. 6 (Dec. 1940): 661; Pasanella, 6. For discussions of this revolt against rationalism and a priori reasoning and the rise of empirical, scientific study of social collectives, see Ross; Morton White, *Social Thought in America: The Revolt against Formalism* (Boston: Beacon Press, 1957); May; Diggins; Louis Menard, *The Metaphysical Club: A Story of Ideas in America* (New York: Farrar, Straus, and Giroux, 2001).

23. Paul Lazarsfeld, "From Technical to Social Knowledge," memo to Hadley Cantril and Frank Stanton, Jan. 1, 1938, Folder 11, Box 26, Series I, Lazarsfeld Collection, 2; Paul Lazarsfeld, "Radio Research and Applied Psychology," *Journal of Applied Psychology* 23, no. 1 (Feb. 1939): 1; Lazarsfeld, "Remarks on Administrative and Critical Communications Research," 2–3. To an organization funding social research such as the Rockefeller Foundation this facade of neutrality that Lazarsfeld's scientific approach offered was appealing. Ross, 400.

24. Lazarsfeld, *Radio and the Printed Page*, xiii, 5–6, 21; H. M. Beville, Jr., *Social Stratification of the Radio Audience* (Princeton, NJ: Princeton Radio Research Project, 1939), v; William Robinson, "Radio Comes to the Farmer," *Radio Research 1941*, ed. Paul Lazarsfeld and Frank Stanton (New York: Duell, Sloan and Pearce, 1941), 249; Hazel Gaudet, "The Role of Radio among Lower Class Urban People," memo to Paul Lazarsfeld, Folder 9, B0080, Bureau of Applied Social Research Collection, Lehman Library, Columbia University; "Proposal for Continuation of Radio Research Project."

25. Cantril and Allport, 36, 270–271; Siepmann, 105; Robinson, 269–272; Herta Herzog, "On Borrowed Experience: An Analysis of Listening to Daytime Sketches," *Studies in Philosophy and Social Science* 9, no. 1 (1941): 86–87; Lazarsfeld, "The Effects of Radio on Public Opinion," 66–70.

26. Lazarsfeld, *Radio and the Printed Page*, 106–108; Suchman, 165–167.

27. John Dewey, for instance, wrote about the importance of educational contexts nearly four decades before Lazarsfeld. John Dewey, *The School and Society and The Child and the Curriculum* (Chicago: University of Chicago Press, 1990), 9–29, 182–209.

28. Lazarsfeld, *Radio and the Printed Page*, 113–120; Lazarsfeld, "The Effects of Radio on Public Opinion," 78; Suchman, 169; Cantril and Allport, 271.

29. Lazarsfeld, *Radio and the Printed Page*, 93; see also Frank, 122.

30. *Co-operative Analysis of Broadcasting: March–June 1930.* pamphlet (New York: Crossley, 1930), Folder Rating Services—CAB, Vertical Files, Broadcast Pioneers Library of American Broadcasting (hereafter BPLAB), University of Maryland, College Park, MD; Erik Barnouw, *A History of Broadcasting in the United States*, vol. 1, *A Tower in Babel; to 1933*, (New York: Oxford University Press, 1966), 270; Herman Hettinger, "Study of Habits and Preferences of Radio Listeners in Philadelphia," 1930, Folder Audience Surveys (2), Box 6, Vertical Files, BPLAB; Herman Hettinger, *A Decade of Radio Advertising* (Chicago: University of Chicago Press, 1933), vi; Rensis Likert, "Method for Measuring the Sales Influence of a Radio Program," *Journal of Applied Psychology* 20, no. 2 (Apr. 1936): 175–182; H. P. Longstaff, "Effectiveness of Children's Radio Programs," *Journal of Applied Psychology* 20, no. 2 (Apr. 1936): 208–220; Harold Gaskill and Richard Holcomb, "The Effectiveness of Appeal in Radio Advertising," *Journal of Applied Psychology* 20, no. 3 (June 1936).

31. Frank Stanton, "The Outlook for Listener Research," Folder 6, Box 3A, Series I, Lazarsfeld Collection, 5–6; Beville, viii; Everette Dennis in Pasanella, 2.

32. Lazarsfeld, "Remarks on Administrative and Critical Communications Research," 2–3; Herman Hettinger and Walter Neff, *Practical Radio Advertising* (New York: Prentice-Hall, 1938), v.

33. Hettinger, *A Decade of Radio Advertising*, vii–viii; Likert, 175; see also Longstaff, 208; Hettinger and Neff, v.

34. Hettinger, *A Decade of Radio Advertising*, vii–viii, 289–290.

35. CBS president William Paley, NBC president Merlin Aylesworth, and Roy Durstine, vice president of one of the nation's leading advertising firms, BBD&O, all contributed to the first volume Hettinger edited, for instance. See *Annals of the American Academy of Political and Social Sciences* 177 (Jan. 1935), *Radio: The Fifth Estate*, ed. Herman Hettinger.

36. Herman Hettinger, "Broadcasting in the United States," *The Annals of the American Academy of Political and Social Sciences* 177 (Jan. 1935), *Radio: The Fifth Estate*, ed. Herman Hettinger, 4, 11.

37. Hettinger, "Broadcasting in the United States," 5, 11; Hettinger, *A Decade of Radio Advertising*, 308.

38. Hettinger, "Broadcasting in the United States," 11.

39. Hettinger, "Broadcasting in the United States," 12.

40. Hettinger, *A Decade of Radio Advertising*, 319–320. This effort to marry laissez-faire liberal capitalism with expert management and greater webs of organization was not entirely new in the Depression. As historian Alan Dawley makes clear, this managerial liberalism expanded on Frederick Taylor's thinking in the early decades of the century, and Herbert Hoover championed this outlook through the 1920s. Alan Dawley, *Struggles for Justice: Social Responsibility and the Liberal State* (Cambridge, MA: Belknap Press of Harvard University Press, 1991), 159–160, 163–165, 319, 344, 347–351.

41. As this implies, Hettinger—and Lazarsfeld as well—played important roles in the maturation of the young field of marketing. Between roughly 1900 and 1920, marketing and business as academic disciplines were born. With the advent of radio advertising, market research took on still greater importance, as businesses sought to understand the impact of a medium with an unseen circulation. Leach, 155–163; Czitrom, 125–126; Dennis in Pasanella, 1–2.

42. David Morrison, "Kultur and Culture: The Case of Theodor W. Adorno and Paul F. Lazarsfeld," *Social Research* 45, no. 2 (Summer 1978): 334, 337, 340, 352, 355; Max Horkheimer, "Preface," *Studies in Philosophy and Social Science* 9, no. 1 (1941), 1 ; Lazarsfeld, "Remarks on Administrative and Critical Communications Research," 14–16; Theodor Adorno, "Memorandum to Dr. Lazarsfeld from T. W. Adorno," Folder 2, Box 25, Series I, Lazarsfeld Collection; Paul Lazarsfeld to Theodor Adorno, no date, Folder 3, Box 25, Series I, Lazarsfeld Collection; Lazarsfeld, "An Episode in the History of Social Research: A Memoir," 323–325.

43. Lazarsfeld, "Remarks on Administrative and Critical Communications Research," 9–10; Czitrom, 142–144.

44. Lazarsfeld to Adorno, 1.

45. Adorno, "Memorandum to Dr. Lazarsfeld from T. W. Adorno," 2; Theodor Adorno, "Lecture Delivered by Theodor Wiesengrund Adorno, January, 1939," B0072, Bureau of Applied Social Research Collection; Theodor Adorno, "Social and Psychological Implications of 'the Radio Voice,'" Office of Radio Research office report, Folder 5, Bureau of Applied Social Research; Morrison, 339; Lazarsfeld to Adorno, 1.

46. Adorno, "Memorandum to Dr. Lazarsfeld from T. W. Adorno," 1.

47. Adorno, "Lecture Delivered by Theodor Wiesengrund Adorno, January, 1939," 23.

48. Adorno, "Lecture Delivered by Theodor Wiesengrund Adorno, January, 1939," 3, 6, 10–13, 18, 20.

49. Adorno, "Lecture Delivered by Theodor Wiesengrund Adorno, January, 1939," 13, 21, 23.

50. Adorno, "Lecture Delivered by Theodor Wiesengrund Adorno, January, 1939," 10; Adorno, "Social and Psychological Implications of 'the Radio Voice,'" 1–2, 6.

51. Adorno, "Social and Psychological Implications of 'the Radio Voice,'" 6; Siepmann, 112.

52. Adorno, "Social and Psychological Implications of 'the Radio Voice,'" 10.

53. Adorno, "Lecture Delivered by Theodor Wiesengrund Adorno, January, 1939," 17–19; Siepmann, 106–113.

54. Adorno, "Social and Psychological Implications of 'the Radio Voice,'" 9–12.

55. Adorno, "Lecture Delivered by Theodor Wiesengrund Adorno, January, 1939," 22; Adorno, "Social and Psychological Implications of 'the Radio Voice,'" 14.

56. Czitrom, 142–144; Dennis in Pasanella, 2; Paul Gorman, *Left Intellectuals and Popular Culture in Twentieth-Century America* (Chapel Hill: University of North Carolina Press, 1996), 176–179.

57. Later communications scholars would level this critique against Lazarsfeld. See Dennis, preface to Pasanella, 2.

Chapter 6

1. William Merrick, *Forgot in the Rains* (broadcast Jan. 9, 1939), in *Columbia Workshop Plays: Fourteen Radio Dramas*, ed. Douglas Coulter (New York: Whittlesey House, 1939), 221–222.

2. John Dos Passos, "The Writer as Technician," in *John Dos Passos: The Major Nonfictional Prose*, ed. Donald Pizer (Detroit: Wayne State University Press, 1988), 169; Norman Corwin, *Seems Radio Is Here to Stay* (broadcast April 24, 1939), in Norman Corwin, *Thirteen by Corwin* (New York: Henry Holt and Company, 1942), 219.

3. Norman Corwin quoted in R. LeRoy Bannerman, *Norman Corwin and Radio: The Golden Years* (University: University of Alabama Press, 1986), 3. See also J. Fred MacDonald, *Don't Touch that Dial! Radio Programming in American Life from 1920 to 1960* (Chicago: Nelson-Hall, 1979), 39, 60; Erik Barnouw, *A History of Broadcasting in the United States*, vol. 2, *The Golden Web: 1933–1953* (New York: Oxford University Press, 1968), 55, 62–64; Glenn Novak, "A Poet's Gift to Radio: Archibald MacLeish's *The Fall of the City*," *West Georgia College Review* 19 (1987): 1; Coulter, vi.

4. Barnouw, 26–27, 62–63; Erik Barnouw in Bannerman, x–xi; Robert McChesney, *Telecommunications, Mass Media, and Democracy: The Battle for the Control of U.S. Broadcasting, 1928–1935* (New York: Oxford University Press, 1993), 188; Laurence Bergeen, *Look Now, Pay Later: The Rise of Network Broadcasting* (Garden City, NY: Doubleday and Company, 1980), 84; Susan Smulyan, *Selling Radio: The Commercialization of American Broadcasting* (Washington, DC: Smithsonian Institution Press, 1994), 142–153; Michele Hilmes, *Radio Voices: American Broadcasting, 1922–1952* (Minneapolis: University of Minnesota Press, 1997), 188–189; Judith Smith, "Radio's 'Cultural Front,' 1938–1948," in *Radio Reader: Essays in the Cultural History of Radio*, ed. Michele Hilmes and Jason Loviglio (New York: Routledge, 2002), 213.

5. Coulter, vi–ix; Barnouw, 63–66; MacDonald, 54–55; Bannerman, 3; Bergeen, 82–83; Novak, 1; Max Wylie in *Fighting Words*, Donald Ogden Stewart, ed. (New York:

Harcourt, Brace and Company, 1940), 82–83; Merrill Denison, "Radio and the Writer," *Theatre Arts Monthly*, 22, no. 5 (May 1938): 369–370.

6. Denison, 368; Archibald MacLeish, *The Fall of the City* (broadcast April 11, 1937), in *Six Plays* (Boston: Houghton Mifflin Company, 1980), 69–93; Novak, 1–8; Barnouw, 66–69; "Radio: Air Raid," *Time Magazine* 32, no. 19 (Oct. 31, 1938): 30. MacLeish's radio play helped give rise to another important radio phenomenon: Orson Welles. Already a successful radio actor, Welles climbed to greater heights when he starred in *The Fall of the City*. A little over a year later, Welles ran his own radio theater company on CBS. Perhaps Welles remembered his experience with *The Fall of the City*'s newscaster-as-narrator device when he used a similar technique in *War of the Worlds* in October 1938.

7. Denison, 366; "Theatre: 'Fall of the City,'" *Time Magazine* 29, no. 16 (April 19, 1937): 60. Coulter, ix. See also Novak, 2, 7–8; Barnouw, 69; Bergeen, 83; Hermine Rich Isaacs, "The Fall of Another City," *Theatre Arts Monthly* 23, no. 2 (Feb. 1939): 149; "Radio: Air Raid," 30.

8. Barnouw, 71–73; MacDonald, 56–58; Lewis Titterton, forward to Arch Oboler, *Fourteen Radio Plays* (New York: Random House, 1940), viii–ix; Robert Landry, introduction to Arch Oboler, *This Freedom: Thirteen New Radio Plays* (New York: Random House, 1942), viii. Media historian Erik Barnouw's description of Oboler's strategy for aurally depicting a body turned inside-out reveals the writer's creativity: "For flesh sounds, a technician put his arm in a length of inner tube, grasped the end firmly, and yanked; for blood-and-guts sounds, warm spaghetti was attacked with bathroom plunger; for bones crunching, Lifesavers were ground between the teeth, very close to the microphone."

9. Carl Van Doren, "Preface," in Corwin, *Thirteen by Corwin*, vii; Bannerman, 20–72; Barnouw, 116–120; MacDonald, 58–60; Norman Corwin to William Lewis, Sept. 13, 1940, in *Norman Corwin's Letters*, A. J. Langguth, ed. (New York: Barricade Books, 1994), 51; Norman Corwin to Davidson Taylor, Jan. 16, 1941, in *Norman Corwin's Letters*, 56.

10. Clifton Fadiman, introduction to Norman Corwin, *More by Corwin: 16 Radio Dramas by Norman Corwin* (New York: Henry Holt and Company, 1944), ix.

11. For critical assessments of radio as an art form, see Jerrold Zinnamon, "Norman Corwin: A Study of Selected Radio Plays by the Noted Author and Dramatist" (PhD dissertation, University of Oregon, 1984), 314; Arthur Miller to Erik Barnouw, April 19, 1945, Folder 20, Box 1, Erik Barnouw Papers, State Historical Society of Wisconsin, Madison. For views of critics and writers who saw a nascent art on the air, see Alan Lomax in Michael Denning, *The Cultural Front: The Laboring of American Culture in the Twentieth Century* (London: Verso, 1996), 91; Zinnamon 312–313; Bannerman, 43–44; Van Doren, vii–ix; Titterton, vii–ix; Landry, vii–ix; Fadiman, ix–xi; Corwin to Lewis, 51; Corwin to Taylor, 56; Norman Corwin to Joan Alexander, Jan. 12, 1941, in *Norman Corwin's Letters*, 53; Coulter, ix–xv; Norman Corwin in *Fighting Words*, 86–90; Norman Corwin, "The Sovereign Word: Some Notes on Radio Drama," *Theatre Arts* 24, no. 2 (Feb. 1940): 130; Arch Oboler, *Oboler Omnibus: Radio Plays and Personalities* (New York: Duell, Sloan and Pearce, 1945), 20; Evan Roberts in *Fighting Words*, 105.

12. For several accounts praising radio's sense of purpose in the late 1930s and early 1940s, see Barnouw, 69–73; MacDonald 59–60; Bannerman xi, 2–5. For counters to this view, see Barnouw, 69; Bergeen, 82; Oboler, *Oboler Omnibus*, 157; Davidson Taylor to Maxim Lieber, Mar. 22, 1940, Box 43; and Langston Hughes to Erik Barnouw, Mar. 27, 1945, Erik Barnouw Folder, Box 10; both in General Correspondence, Langston Hughes Papers, James Weldon Johnson Collection in the Yale Collection of American Literature,

Beinecke Rare Book and Manuscript Library, New Haven, Connecticut. The play in question was Oboler's *This Precious Freedom*. For a discussion of progressives on the air and the decline of those opportunities, see Smith, "Radio's 'Cultural Front,' 1938–1948," 209–225.

13. Oboler, *Oboler Omnibus*, 20.

14. Archibald MacLeish, "Public Speech and Private Speech in Poetry," *Yale Review* 27, no. 3 (March 1938): 541, 546–547.

15. Laura Browder, *Rousing the Nation: Radical Culture in Depression America* (Amherst: University of Massachusetts Press, 1998), 3–5; James Oppenheim, "The Plight of the Poet in the Machine Age," in *Behold America!* ed. Samuel Schmalhausen (New York: Farrar and Rinehart, 1931), 462–472. Despite Oppenheim's gloomy assessment, historian Terry Cooney suggests that *Seven Arts* proved influential: the journal, Cooney says, argued that art absolutely had to tackle public issues and criticized writers who failed to do so. Terry Cooney, *The Rise of the New York Intellectuals:* Partisan Review *and Its Circle* (Madison: University of Wisconsin Press, 1986), 26–27.

16. A. Joan Saab, *For the Millions: American Art and Culture Between the Wars* (Philadelphia: University of Pennsylvania Press, 2004); Robert McElvaine, *The Great Depression: America, 1929–1941* (New York: Times Books, 1993), 268–275; Denning, xvii, 365–375; Karal Ann Marling, *Wall-to-Wall America: A Cultural History of Post-Office Murals in the Great Depression* (Minneapolis: University of Minnesota Press, 1982), 7; Browder, 7; Browder, 12–13; Roger Hill and Orson Welles, *Everybody's Shakespeare: Three Plays Edited for Reading and Arranged for Staging* (Woodstock, IL: The Todd Press, 1934).

17. Denning, throughout, but see particularly xvi–xviii, 4–50, 60, 122; Cooney, 28–29, 71, 129–147, 204–206; Browder, 7–13; Lary May, *The Big Tomorrow: Hollywood and the Politics of the American Way* (Chicago: University of Chicago Press, 2000), 1–53; Richard Pells, *Radical Visions and American Dreams: Culture and Social Thought in the Depression Years* (New York: Harper and Row, Publishers, 1973), 152, 158; Dos Passos, 169–172.

18. Norman Corwin, *Appointment* (broadcast June 1, 1941), in *Thirteen by Corwin*, 297; Archibald MacLeish in Barnouw, 68–69.

19. Corwin, *Seems Radio Is Here to Stay*, 222, 224; Corwin in *Fighting Words*, 86–87; Norman Corwin, *Years of the Electric Ear: Norman Corwin; A Directors Guild of America Oral History*, interview by Douglas Bell (Metuchen, NJ: Directors Guild of America, 1994), 25; Zinnamon, 327; Oboler, *Fourteen Radio Plays*, xv, xxix.

20. Corwin, *Appointment*, 312; Arch Oboler, *Bathysphere* (broadcast Sept. 1939), in *Fourteen Radio Plays*, 57–76; MacLeish, *The Fall of the City*, 91–93.

21. Corwin in *Fighting Words*, 87.

22. Corwin, "The Sovereign Word," 130–131; Archibald MacLeish in Denison, 367; Arch Oboler in *Fighting Words*, 98. See also Corwin in *Fighting Words*, 87; Oboler, *This Freedom*, 141; Norman Corwin in Amy Bonner, "Verse Drama and the Radio: A Market Survey and Technical Guide," *Poetry: A Magazine of Verse*, 58, no. 5 (Aug. 1941): 283.

23. Arch Oboler, *Dark World*, in *This Freedom*, 74; Norman Corwin to Frank Farrara, March 20, 1945, in *Norman Corwin's Letters*.

24. As Denning suggests, radio was a site of some of the Left's most vocal success, but remains largely unexplored by historians. Although he focuses his attention on Orson Welles as an example of the movement's influence on radio artists, he also briefly mentions MacLeish, Corwin, and *Columbia Workshop* producer Irving Reis as deeply connected to the popular front. Denning, throughout, but see particularly xvi–xviii, 4–50, 60,

91, 122, 362–383; Cooney, 28–29, 71, 129–147, 204–206; Browder, 8–13; Pells, 111–112, 152–158; May, 3–99; Raymond Gram Swing, *Forerunners of American Fascism* (Freeport, NY: Books for Libraries Press, 1969; orig. 1935), 22.

25. Arch Oboler, *Mr. Whiskers*, in *Fourteen Radio Plays*, 143. The judge at the immigrant shopkeeper's citizenship hearing was so impressed with the shopkeeper's understanding of American values that the judge awarded the immigrant citizenship despite his protests. For other examples of Oboler's assaults on the unbridled pursuit of profits see, for instance, Oboler, *Profits Unlimited*, in *Fourteen Radio Plays*, 217–236; Oboler, *Mr. Ginsberg*, in *Fourteen Radio Plays*, 197–212.

26. Arch Oboler, *This Precious Freedom* (broadcast Dec. 30, 1939), in *Fourteen Radio Plays*, 147–162; Oboler, *Bathysphere*, 72–73. For a similar contemporary view of fascism, see Swing, 22.

27. James Boyd, ed., *The Free Company Presents . . . A Collection of Plays about the Meaning of America* (New York: Dodd, Mead and Company, 1941). CBS ran the Free Company's plays in early 1941. Once the United States entered World War II, of course, networks became still more willing to lend their services to oppose fascism.

28. MacLeish, *The Fall of the City*, 69–93; Archibald MacLeish, *Air Raid* (broadcast in 1938), in *Six Plays*, 99–123; Norman Corwin, *They Fly through the Air with the Greatest of Ease* (broadcast Feb. 19, 1939), in *Thirteen by Corwin*, 57–77, quote on p. 68.

29. Arch Oboler, *And Adam Begot*, in *This Freedom*, 49–68.

30. Arch Oboler, *The Man to Hate* (broadcast Aug. 5, 1939), in *Fourteen Radio Plays*, 112–119; Arch Oboler, *This Precious Freedom* (broadcast in 1940), in *Oboler Omnibus*, 159–176; Norman Corwin, *The Oracle of Philadelphia* (broadcast April 21, 1940), in *Thirteen by Corwin*, 324.

31. Norman Corwin to Joan Alexander, Jan. 12, 1941, *Norman Corwin's Letters*; Corwin, *Seems Radio Is Here to Stay*, 218.

32. Arch Oboler, *The Laughing Man* (broadcast June 3, 1939), in *Fourteen Radio Plays*, 3–6. It is worth remembering here that in the late 1930s many used the term "race" to refer to separate nationalities and certainly to refer to Jewishness. In 1939, then, Oboler's commentary applied most evidently to the situation in Germany—though he may have also intended it as a less direct critique of the United States too.

33. Corwin, *Seems Radio Is Here to Stay*, 236; Arch Oboler, *The Word*, in *Oboler Omnibus*, 68–86. For other, less overt, examples of plays focusing on what seemingly different people shared see Arch Oboler, *Mr. Whiskers*; Norman Corwin, *Daybreak* (broadcast June 22, 1941), in *Thirteen by Corwin*, 117–133. The former emphasizes that diverse ethnic groups share a vital Americanness; the latter dwells upon the basic common experiences that span the globe.

34. Norman Corwin, *Between Americans* (broadcast July 6, 1941), in *More by Corwin*, 342.

35. There was a wide variety in quality—no doubt because even the likes of a Corwin or an Oboler had a regular weekly deadline they had to meet. The schedule was, as Oboler saw it, something of a grind: "Radio, for the dramatist, is a huge, insatiable sausage grinder into which he feeds his creative life," he said. Arch Oboler quoted in MacDonald, 58.

36. Denning, 375–377.

37. Oboler, *Oboler Omnibus*, 223.

38. For a discussion of some of the intellectual underpinnings of Modernist thought, consider Daniel Singal, "Towards a Definition of American Modernism," *American Quarterly* 39, no. 1 (Spring 1987): 9–19; John Dewey, *The Influence of Darwinism on Philosophy* (New York: Henry Holt and Co., 1910), 3; Morton White, *Social Thought in America: The Revolt against Formalism* (Boston: Beacon Press, 1957), 11–15; John Diggins, *The Promise of Pragmatism: Modernism and the Crisis of Knowledge and Authority* (Chicago: University of Chicago Press, 1994), 7–8, 124–127; Henry May, *The End of American Innocence: A Study of the First Years of Our Time* (Chicago: Quadrangle Books, 1964), 140–165; Louis Menard, *The Metaphysical Club: A Story of Ideas in America* (New York: Farrar, Straus and Giroux, 2001); Henry James, *The Real Thing* (London: Macmillan and Co., 1893), 1–41.

39. Historian A. Joan Saab has described the leading artistic movement of the Depression as "a participatory form of American modernism." Saab, 11.

40. Irving Reis, *Meridian 7-1212*, in Coulter, ed., 33–62.

41. Corwin, "The Sovereign Word," 133; Corwin, *Seems Radio Is Here to Stay*, 217; Corwin, *They Fly through the Air with the Greatest of Ease*, 59–73. See also Zinnamon, 65–66.

42. Corwin, "The Sovereign Word," 134. See also Novak, 3; "Radio: Air Raid," 30.

43. MacLeish, *The Fall of the City*, 73; for a discussion of Welles's intent, see Robert Brown, *Manipulating the Ether: The Power of Broadcast Radio in Thirties America* (Jefferson, NC: McFarland and Company, 1998), 227.

44. Peter Conn, *The Divided Mind: Ideology and Imagination in America, 1898–1917* (Cambridge: Cambridge University Press, 1983), 282–283.

45. Corwin, *Years of the Electric Ear*, 71; Oboler, *Fourteen Radio Plays*, xv. Oboler so delighted in the possibility of using radio to leap into the past or the future that he felt he should have a crest featuring a dinosaur standing among the stars. Barnouw, 72.

46. MacLeish, *Air Raid*, 99–101; Norman Corwin, *The Plot to Overthrow Christmas* (broadcast Dec. 25, 1938), in *Thirteen by Corwin*, 89–90, 103–104; Norman Corwin, *The Odyssey of Runyon Jones* (broadcast June 8, 1941), in *Thirteen by Corwin*, 3–22; Norman Corwin, *Anatomy of Sound* (broadcast Sept. 7, 1941), in *More by Corwin*, 244–245.

47. Corwin, *Daybreak*, 117, 123.

48. Vic Knight, *Cartwheel*, in Coulter, ed., 247–259; Arch Oboler, *The Man to Hate* (broadcast Aug. 5, 1939), in *Fourteen Radio Plays*, 113–118. See also Arch Oboler, *This Lonely Heart* (broadcast Aug. 29, 1939), in *Fourteen Radio Plays*, 8–46; Arch Oboler, *And Adam Begot*, in *This Freedom*, 49–68. Corwin too leapt through time occasionally. In *To Tim at Twenty*, for example, Corwin had one of his characters consciously try to bring two periods together: the protagonist, a flyer for the British air force sent on an almost-certain suicide mission, tried to relate to his then-young son as an adult by writing a letter to him to be opened years in the future. Norman Corwin, *To Tim at Twenty* (broadcast Aug. 19, 1940), in *Thirteen by Corwin*, 249–257.

49. Van Doren, viii; Zinnamon, 313; Titterton, viii; MacDonald, 57.

50. Corwin, *Appointment*, 301–305; Corwin, *Daybreak*, 118–128; Oboler, *This Precious Freedom*, in *Fourteen Radio Plays*, 150; Corwin in Zinnamon, 313.

51. According to radio's most prominent writers, in its use of sound radio outshone visual media such as film—at least prior to Welles's work. In film and other visual arts, MacLeish and Corwin argued in 1938 and 1940, the image overwhelmed the word and

sound, limiting the effectiveness of the expression in important regards. MacLeish in Denison, 367; Corwin, "The Sovereign Word," 130–131.

52. Coulter, vii.

53. Arch Oboler, *Mr. Ginsburg*, in *Fourteen Radio Plays*, 211–212. Arch Oboler, *The Ugliest Man in the World*, in *Oboler Omnibus*, 49. For examples of Oboler plays in which the narratives slipped into the past following his speakers' thoughts see Oboler, *The Man to Hate*, 112–118; Oboler, *This Lonely Heart*, 8–46. According to Oboler, NBC's Wyliss Cooper deserved credit for originating the stream-of-consciousness form on radio. Arch Oboler to Erik Barnouw, May 13, 1945, Folder 22, Box 1, Erik Barnouw Papers.

54. Arch Oboler to Erik Barnouw. Indeed, by his own estimate, Oboler used the stream-of-consciousness technique in only about one in ten plays. But his success with his expressionistic dramas ensured that his protestations would not shake his reputation.

55. Corwin, *Descent of the Gods* (broadcast August 31, 1941), in *More by Corwin*, 111–116.

56. Norman Corwin, *Mary and the Fairy* (broadcast August 3, 1941), in *More by Corwin*, 3–21.

57. Oboler, *Oboler Omnibus*, 18–20, 179–180; Oboler, *This Freedom*, 48; Norman Corwin, *Radio Primer* (broadcast May 4, 1941), *Thirteen by Corwin*, 31; Norman Corwin to Davidson Taylor, Feb. 21, 1941, *Norman Corwin's Letters*, 59. See also Norman Corwin to LeRoy Bannerman, Oct. 26, 1977, *Norman Corwin's Letters*, 344.

58. Smith, 217–225; Barnouw, 151–154, 163, 253–283; MacDonald, 58–60, 65–68, 76–86; Corwin, *Years of the Electric Ear*, 83–89; Bannerman, 9–10, 187–188, 199–201; Bergeen, 131–136.

59. Corwin in *Fighting Words*, 90.

60. Skulda Baner, Milwaukee, to FCC, Oct. 31, 1938, Box 237, Research Group 173, FCC Collection, National Archives, College Park, MD.

Conclusion

1. CBS press release, Jan. 25, 1937; and George Patterson, WAVE program director, to John Royal, Feb. 24, 1937; both in Folder 35, Box 53, NBC Papers, State Historical Society of Wisconsin, Madison, WI (hereafter NBC Papers); "Mayor Urges City Residents to Keep Calm," *Louisville Courier-Journal*, Jan. 22, 1937, 1; "WHAS Radiocasts Flood Warnings," *Louisville Courier-Journal*, Jan. 22, 1937, 3; "Attention Rescue Workers," *Louisville Courier-Journal*, Jan. 24, 1937, 1; "Boat Patrols Assisting in Relief Tasks," *Louisville Courier-Journal*, Jan. 24, 1937, 5; Herman Landau, "News, Radio," *Louisville Courier-Journal*, Jan. 24, 1937, Magazine section, 3; Robert Brown, *Manipulating the Ether: The Power of Broadcast Radio in Thirties America* (Jefferson, NC: McFarland and Company, 1998), 139. Anne Clatfelter, Attalla, Alabama, to NBC, Jan. 26, 1937; Capt. Robert Foster to NBC, Jan. 27, 1937; and William Miller to John Royal, internal memo, Jan. 30, 1937, all in Folder 35, Box 53, NBC Papers; "Radio Stations Raising Funds to Aid Victims," *Louisville Courier-Journal/Louisville Times*, Jan. 26, 1937, 1; "Mayor Miller Reviews Flood Crisis in National Radiocast," *Louisville Courier-Journal/Louisville Times*, Jan. 27, 1937, 3; "Louisville Carries On," *Louisville Courier-Journal*, Jan. 21, 1937, 1.

2. For brief discussions of media and the civil rights movement, see Julian Bond, "The Media and the Movement: Looking Back from the Southern Front," in *Media, Culture, and the Modern African American Freedom Struggle*, ed. Brian Ward (Gainesville: University

of Florida Press, 2001), 16–40, especially 31; Brian Ward, *Radio and the Struggle for Civil Rights in the South* (Gainesville: University Press of Florida, 2004), 4–5. For a turn-of-the-century reflection on the democratic importance of a mass audience and common culture, see Cass Sunstein, *Republic.com* (Princeton, NJ: Princeton University Press, 2001). It should be emphasized that Sunstein does not argue against diversity or weigh in on the culture wars—as so many commentators calling for a common American culture did in the late twentieth century. Rather he accepts heterogeneity, but suggests that a form of general-interest communication is essential in order to provide common ground amidst that welcome diversity.

3. Will Rogers, NBC anniversary program address, November, 1933, Folder 33, Box, 21, NBC Papers.

4. Jane Ziegler, Saugerties, New Jersey, to H. V. Kaltenborn, June 24, 1930, Folder June-Oct. 1930, Box 17, H. V. Kaltenborn Papers, State Historical Society of Wisconsin.